W9-COT-391

Steven A. Edwards

The Nanotech Pioneers

Also of Interest

Renn, J. (ed.)

Albert Einstein – Chief Engineer of the Universe
Exhibition Catalogue and Documents

725 pages in 2 volumes
2005, Hardcover
ISBN 3-527-40571-2

Borisenko, V. E., Ossicini, S.

What is What in the Nanoworld
A Handbook on Nanoscience and Nanotechnology

347 pages with 120 figures and 28 tables
2004, Hardcover
ISBN 3-527-40493-7

Huebener, R.

Electrons in Action
Roads to Modern Computers and Electronics

227 pages with 86 figures and 1 table
2005, Hardcover
ISBN 3-527-40443-0

Köhler, M., Fritzsche, W.

Nanotechnology
An Introduction to Nanostructuring Techniques

284 pages with 143 figures and 9 tables
2004, Hardcover
ISBN 3-527-30750-8

Fecht, H.-J., Werner, M. (eds.)

The Nano-Micro Interface
Bridging the Micro and Nano Worlds

351 pages with 102 figures and 27 tables
2004, Hardcover
ISBN 3-527-30978-0

Ajayan, P. M., Schadler, L. S., Braun, P. V.

Nanocomposite Science and Technology

239 pages with 126 figures and 5 tables
2003, Hardcover
ISBN 3-527-30359-6

Steven A. Edwards

The Nanotech Pioneers

Where Are They Taking Us?

WILEY-VCH Verlag GmbH & Co. KGaA

The Author

Steven A. Edwards
S. A. Edwards & Assoc., Christiana, USA
SAlanEd@aol.com

Coverpicture
Eric J. Heller, *Nanowire*
Used with permission by Resonance Fine Art,
Eric J. Heller, Cambridge

Library of Congress Card No.: applied for

British Library Cataloguing-in-Publication Data
A catalogue record for this book is available from the British Library.

Bibliographic information published by Die Deutsche Bibliothek
Die Deutsche Bibliothek lists this publication in the Deutsche Nationalbibliografie; detailed bibliographic data is available in the Internet at <http://dnb.ddb.de>.

Printed in the Federal Republic of Germany.
Printed on acid-free paper.

Composition Kühn & Weyh, Satz und Medien, Freiburg
Printing betz-druck GmbH, Darmstadt
Bookbinding Schäffer GmbH, Grünstadt

ISBN-13: 978-3-527-31290-0
ISBN-10: 3-527-31290-0

Contents

The Nanotech Pioneers. Steven A. Edwards

ISBN: 3-527-31290-0

Foreword

Nanotech Pioneers provides an insightful look into the nanotechnology revolution, where it is going, and how it will impact us. And it introduces us to the fascinating cast of characters that are bringing it into existence.

Author Steve Edwards has been on the inside track of nanotechnology, as both a scientist and a journalist. Steve has identified the exciting technologies and intriguing players. Some of them are capturing the headlines, and many others you are not likely to hear about, but you will definitely want to know.

The nanotechnology revolution has similarities to the wild forecasts that accompanied the Internet craze, until the bubble burst, but it is very different. Nanotech effects the material stuff the world is made of, including things that make a lot of money, pharmaceuticals, cosmetics, polymers, precious metals, clothing, cars, fuels, steel, diamonds, DNA, cells, bones, blood, brains, computers, semiconductors, biosensors, computer screens, watches, lasers, space travel, pet food, kitty letter, and much, much more. And in each of these situations, nanotechnology has the potential to make money. A lot of money.

Steve Edwards has been a prominent scientific writer and conference organizer who has covered the many advances in material science and nanotechnology, and how they are affecting industry and commerce. I have been a speaker at many conferences where Steve was in attendance, and often a moderator.

I can recall discussions over lunch, or over coffee, at many of these conferences, with Steve and with the other conference attendees, and networking to keep track of the next major advance in technology, the latest scientific papers, the hot companies attracting venture capital, the people moving up, down, sideways or out of the industry. The big companies shopping for technology and applications, or spinning out companies.

Steve was there, listening to this tempo of the business as major advances in science and technology started spilling out of the labs and into businesses. Steve was taking notes and mapping out the trends, the people, the forecasts, and the cool technologies.

At these meetings appeared venture capitalists, economists and government officials, who debated the size of the emerging market, some saying it would be many billions, and some many trillions, of dollars.

The Nanotech Pioneers. Steven A. Edwards

ISBN: 3-527-31290-0

In the midst of all these high-minded, breathtaking visions of the future, were entrepreneurs looking for capital, scientists looking for entrepreneurs, reporters looking for stories, consultants looking for consultees, and venture capitalists looking to invest.

Steve has, in this book, captured the spirit and the excitement and the intriguing personalities who have come together to create this new world of business and science. *Nanotech Pioneers* is a great way to see how interesting little ideas are turning into companies, some of which are bound to be the next IBM or Microsoft. And others will fade into obscurity.

Not much happens in burgeoning industries without entrepreneurial energy. What I really like about Steve's storytelling is he mixes accurate science with down-to-earth pragmatics and a real skill for describing the people, and the entrepreneurial network. Too many books and articles on nanotechnology are written by people who don't know the science, and therefore just repeat whatever the entrepreneurs are telling them. Steve, who has a Ph.D. in molecular biology, can separate the wheat from the chaff, and cull out the hype from the real promise. But you do not have to be at all technical to really enjoy this book. Whether or not you are technical, you will be exposed to many of the really cool technologies that are being enabled and impacted by nanotech.

It is interesting how inadvertency plays a role in thrusting people onto the scene of nanotechnology. This happened to me, and Steve has done a great job of capturing how a little thing can change your life and change an industry. I am the CEO of a company, Biophan, and a few years ago a scientist told me that nanotubes could solve some of the problems my company was seeking to solve. The next thing I knew, my company became one of the more talked about and successful nanotech companies. And just a few years later, I was sitting in a conference, listening to a speaker talk about the discovery of tiny little microtubules inside of a deposit of clay in a mine in Utah and, having gotten into a project using nanotechnology, I was attuned to the potential for nanotubes. They are usually made of carbon and created in labs. But here was a fellow describing a naturally occurring nanotube.

Sitting there, in that conference, I could envision dozens of uses and could see how a company could be formed to develop the means to separate the tubes from clay, and apply them to many new applications. I mentioned this to Steve, and explored the situation in depth. Now there is a public company, NaturalNano, pursuing this in earnest. I am an investor, and on the board of NaturalNano. And I am the subject of one of the vignettes in this book. Explaining how this all came about, inadvertently. And that is one of the things I really like about *Nanotech Pioneers*. Steve was there, as this industry has emerged, and watched many of the players reported on in this book run the gauntlet from start-up to success.

But beware, there are some hazards to reading further. The nanotech revolution is contagious, and there is risk you can get roped in! I recommend you hold this book at least six inches from your soul, because nanotechnology is compelling and contagious, once you get what it is about, and what it can mean to the world and to business.

I am fortunate, having been in the nanotechnology field for the past five years, to have met many of the people Steve reports on in *Nanotech Pioneers*, and to be familiar with many of the companies and technologies.

I am so pleased with the job Steve did, knowing – first hand – how successfully he has captured the facts, the promise, the people and personalities, and the excitement, that is driving this revolution.

I agreed to write this foreword, to encourage you to take this book home with you, or on your next business trip, and find out about something going on right around you, in every town and hamlet with a university of a research lab, that is truly going to change the planet and make many things much, much better – and some people much, much richer.

Michael Weiner
Founder and CEO of Biophan, Inc.
Founder of Natural Nano, Inc.

Acknowledgments

First, foremost and always, I would like to thank my wife Sally, who puts up with me and keeps me sane.

Louis Naturman, president and founder of Business Communications Company (www.bccresearch.com) was the one who really put me on to nanotechnology. At first, I told him it was only a buzzword, but I was wrong. It was as editor of BCC's Nano/Bio Convergence News (now, alas, defunct) that I gained a broad understanding of the capabilities of nanotechnology. I have also helped Lou put on nanotechnology conferences for several years now. It has turned out to be an expensive hobby for him, but it has been to the general good, as the speakers at those conferences will acknowledge. I hope that the world will recognize his contribution.

RedZone Profits, a division of Taipan Group, has let me channel some of the information gained in writing this book into income-generating articles on the nanotech industry, for which I heartily thank them.

I would like particularly to acknowledge the many people within the nanotechnology community who have given me information, either in formal interviews or somewhat unknowingly over lunch at a conference. To the latter, I apologize. I never said I wasn't a journalist.

I would certainly like to thank Martin Ottmar and the staff at Wiley VCH for the opportunity to write this book.

I would be remiss if I did not also thank Larry Page and Sergey Brin, the inventors of the Google search engine, an invaluable research tool. I vaguely remember typing my Ph.D. thesis on an IBM Selectric typewriter before there were word processors, but I don't know how I lived before Google came into being.

Steven A. Edwards, Ph.D.
S.A. Edwards and Associates

The Nanotech Pioneers. Steven A. Edwards

ISBN: 3-527-31290-0

Chapter 1
The Promise of Nanotechnology

A technological journey is underway – a trip into very small spaces. The journey is led by an eclectic band of engineers and scientists from all disciplines – biology, chemistry, physics and mathematics – who are pooling their talents to create a new field called "nanotechnology". The destination of this journey is not yet entirely clear. Are these nanotech pioneers leading us into a new world of bountiful productivity, or into a dangerous ravaged landscape?

When Lewis and Clark set off from St. Louis, Thomas Jefferson gave them a mandate "...to explore the Missouri river and such principal stream of it, as by its course and communication with the waters of the Pacific Ocean, may offer the most direct and practical water communication across this continent, for the purpose of commerce" [1]. When the Manhattan Project was formed under the greatest of secrecy, the purpose was clear to its participants – to create an atomic bomb that would, by its extraordinary power, put an end to the Second World War. When President John F. Kennedy promised to put Americans on the moon within a decade, there was no doubt as to the destination, although we seem to have forgotten what we were going to do once we got there.

Though funded by billions of dollars from governments around the world and billions more from private industry, the nanotech effort has no overarching mission statement. In this gold rush, the miners have hitched up their wagons and are heading out into uncharted territory. The nanotech journey is open-ended. It is as if, halfway through the Cumberland Gap, Daniel Boone had gathered his followers around him and said, "Well, in a few days we are either going to settle Kentucky, take a tour of Disneyland, or grab a space launch to Jupiter."

One focus of the Nanotech Pioneers is clear: they are out to change the way that we build things now with bulk materials, whittling them down or molding them, to a model that is more like that used by living things, creating objects with defined features that extend to the molecular level. Nanotech seeks to "...rebuild the world one molecule (or even one atom) a time", or so the slogan goes. But is the world really in need of rebuilding?

The more extreme nanotech enthusiasts believe that this new technology will usher in a kind of utopia where material goods will self-assemble from elemental feedstocks in the way that seeds turn into flowers. Some observers, paradoxically,

The Nanotech Pioneers. Steven A. Edwards

ISBN: 3-527-31290-0

are concerned that nanotech will usher in such an era of abundance that traditional economics based on scarcity will fail, and that the capitalist system and the social organization it has engendered will be in peril.

Nanotech detractors see the technology as extremely dangerous. Some worry about the "gray goo" scenario wherein runaway nanobots run riot and turn the biosphere into dust, destroying human life in the process. Others worry about more conventional environmental contamination – that nanoparticles might have carcinogenic properties similar to asbestos, for instance.

Not since the early days of the nuclear power industry has there been a wider divergence between proponents and opponents of a new technology. Boosters of nuclear power suggested that electric power would become "... to cheap to meter" and that dependence on fossil fuels would fall by the wayside. Detractors warned that reactors would self-destruct in atomic bomb-like explosions, leaving large swaths of radioactive territory that would be uninhabitable for generations. The truth, of course, has been somewhere in between.

In the coming chapters, we will explore the benefits and opportunities of nanotechnology, as well as its potential dangers.

Defining Nanotechnology

What, actually, do we mean by nanotechnology? The term itself was first coined in 1974 by Tokyo Science University professor Norio Taniguchi, who used it to describe the extension of traditional silicon machining down into regions smaller than one micron. By a now more generally accepted definition, today, nanotechnology is the engineering and fabrication of objects with features smaller than 100 nanometers, or one-tenth of a micron. A micron (μm) is one millionth of a meter – too small for the eye to resolve. A nanometer (nm) is 1 thousandth of a micron – that is to say, really, really tiny. One nanometer is about the size of six carbon atoms aligned, or 10 hydrogen atoms – objects too small to see or image except by the use of very powerful electron or atomic force microscopes. So we are talking about a molecular scale.

The thinnest thing, apparently, that most people are generally aware of is a human hair. So texts and articles on nanotechnology will tell you that a nanometer is 60 000 times smaller than a human hair is in diameter. Or sometimes the number is 100 000; nobody seems to agree. I, personally, have very thin, baby-type hair. In a laboratory long ago, in a place far away, for the purpose of impressing my daughter, I took one of each of our hairs and placed them under a microscope. Her hair looked like a cable compared to mine. So I don't use this hackneyed human hair comparison to give people an idea of nanometers. Human hair varies a lot, OK? And mine is almost gone, anyway.

Look at Table 1, which lists the sizes of some fairly well-known biological objects. A white blood cell is about 10 μm or 10 000 nm in diameter. Note that this is actually larger than the interior diameter of the smallest capillaries (8000 nm), so it helps that blood cells are deformable. Bacteria can be as large as a white blood cell, but most are much smaller, on the order of 1 μm in diameter. Viruses are smaller still, with an upper size range of about 100 nm. Nanofabricated objects have architectural features sizes that are equal to or smaller than the diameter of a virus.

Currently mass-produced semiconductor chips can have circuit elements etched down to 90 nm in diameter. However, this is falling rapidly with new nanolithography techniques, which are already pushing the limit down to around 20 nm, or smaller than the diameter of a ribosome, the organelle within our cells that makes proteins.

Carbon nanotubes (see below) can have diameters smaller than 2 nm – hence their desirability as potential components in nanoscale chips. Another staple of nanotechnology, the quantum dot, can be manufactured reliably as small as 2 nm in diameter. These enigmatic objects have a variety of uses in biosensors and in electronics, as will be discussed in following chapters.

Table 1 The sizes of nanoscale objects: Nature versus fabrication.

Object	Diameter
Hydrogen atom	0.1 nm
Buckminsterfullerene (C60)	0.7 nm
Carbon nanotube (single wall)	0.4–1.8 nm
6 carbon atoms aligned	1 nm
DNA	2 nm
Proteins	5–50 nm
CdSe Quantum Dot	2–10 nm
Ribosome	25 nm
Virus	75–100 nm
Semiconductor Chip Features	90 nm or above
Mitochondria	500–1000 nm
Bacteria	1000–10 000 nm
Capillary (diameter)	8000 nm
White blood cell	10 000 nm

Top-Down versus Bottom-Up Manufacturing

Nanoscale manufacturing can occur either from the "top down" or the from the "bottom up." Top-down manufacturing starts with bulk materials which are then whittled down, until the features that are left are nanoscale. For instance, crystalline drugs may be milled until the individual particle sizes are 100 nm, or smaller. At this size, the particles have a much larger surface area in relation to volume than would more conventional microscale particles. This allows them to dissolve much faster – which is critical for certain drugs that are not very soluble in water.

Bottom-up manufacturing involves creating objects or materials from individual atoms or molecules and then joining them together in a specific fashion.

Think about how a table is built. A plank of wood is connected to three or four posts, through the use of screws and wood-glue. The posts may also be made of wood. Simple enough. This is classical bulk material manufacturing. But how is the wood made?

Wood is created by joining molecule to molecule according to instructions decoded from the DNA in the cells of trees. Tree-trunks may extend hundreds of feet into the air, bringing water from the roots to support branches and leaves. Whole ecosystems that live in the upper reaches of the rainforests are dependent upon this remarkable material. And yet, wood is synthesized at the nanoscale by the individual cells of the tree.

What is the chemical composition of wood? Wood is largely made of cellulose, which is in turn composed of repeating units of glucose, a simple sugar (a single unit is shown in square brackets in Fig. 1). A related material, potato starch, is also composed of repeating units of glucose (Fig. 2). So why can't we build houses out of potatoes? Unlike cellulose, potato starch is not rigid at all. The differences between cellulose and starch reside primarily in the molecular link that connects one glucose units one to another. These links translate into wholly different properties.

This is the promise of nanotechnology – to find extraordinary properties in the arrangement of simple materials.

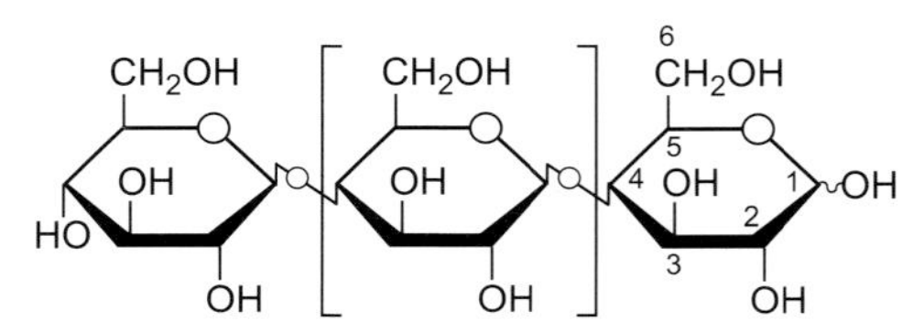

Figure 1 The chemical composition of cellulose. Brackets indicate the boundary of a glucose subunit. The carbon numbering system is indicated in the last subunit to the right.

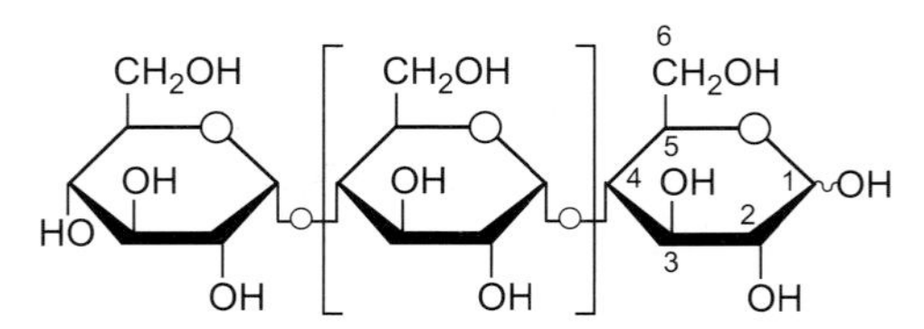

Figure 2 The chemical composition of starch. Note that the only difference between the two structures lies in the placement of the bond between glucose subunits.

Cellulose is composed entirely of carbon, hydrogen, and oxygen, as shown above. Burn a tree trunk and the cellulose will be oxidized to carbon dioxide and water. However, because a fire is rarely completely efficient, the ashes remaining will contain a lot of elemental carbon remaining in the form of soot.

What is in Soot? The Different Forms of Carbon

A component of soot is colloidal carbon, which is also manufactured under more controlled conditions as carbon black. This is a nanoparticle that has been used for centuries as a pigment in inks, paints, and finishes; today, it is also used as a reinforcing agent in rubber, notably in tires. Carbon black is actually small enough that it will enter the skin. Workers at tire factories may sweat out carbon black onto their clothes and sheets for a week or two after they have ceased employment.

Elemental carbon is also used in the form of graphite as a lubricant (Fig. 3), or to make extremely strong carbon fiber material used in bicycles and tennis rackets.

A rare component of soot is a cylindrical form of carbon called a nanotube. Carbon nanotubes can be thought of as a single layer of graphite (called a graphene sheet) rolled into a cylindrical tube. Variants of the structure exist, depending on

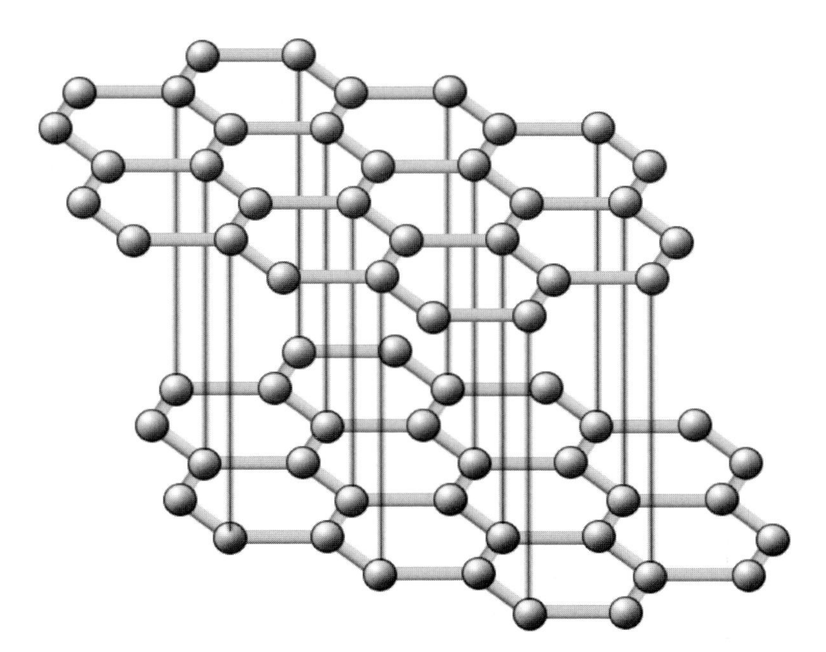

Figure 3 Molecular model of graphite. Each of the atoms is a carbon molecule bound to three other carbon molecules in the same plane. The planar surfaces do not have covalent links and are therefore free to move relative to one another, which gives graphite its lubricant properties. Image reproduced courtesy of Samantha J. Shanley, University of Bristol.

how the ends of the sheet connect and the diameter of the cylinder (Fig. 4). These tubes may or may not have a curved cap on either end. Carbon nanotubes are many times stronger than steel, and conduct electricity better than copper – as will be discussed in a later chapter. Carbon nanotubes have become iconic devices for the field of nanotechnology. Small companies are now in the process of developing nanotubes into transistors and memory devices for computers. It is expected that ton quantities of carbon nanotubes will be produced annually within a few years.

Under extreme pressure, elemental carbon will also spontaneously form into a very different crystalline form called a diamond (Fig. 5). Diamond is the densest form of carbon, packing the most atoms into the smallest area. (Next time you see a multi-karat chunk of diamond on somebody's ring, don't get jealous; just remind yourself that it's basically a hunk of very compressed charcoal).

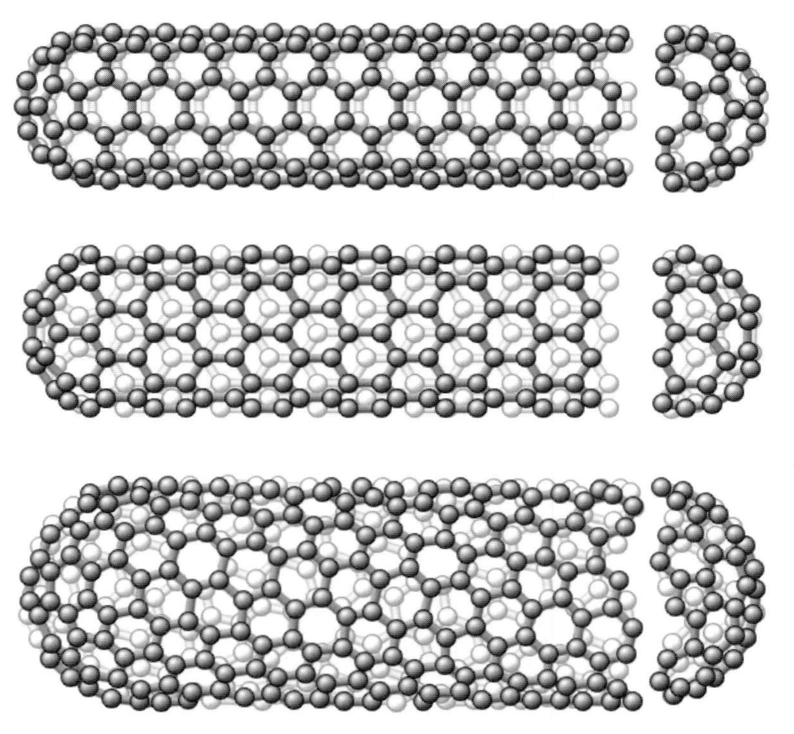

Figure 4 Carbon nanotube structures. Each carbon is bound to each other in a cylindrical arrangement. These may be thought of as graphite planes that have been cut and rolled up. Slightly different arrangements occur, depending upon how the sheets are cut and the diameter of the tube. Tubes may or may not have a cap at either end. Images reproduced courtesy of Samantha J. Shanley, University of Bristol.

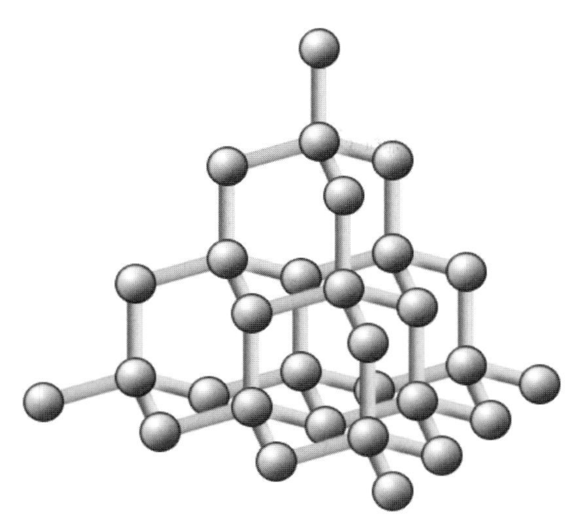

Figure 5 Molecular model of diamond. Each carbon atom is bound to three others in a three-dimensional crystal. Image reproduced courtesy of Samantha J. Shanley, University of Bristol.

An even more striking version of carbon is a molecule called buckminsterfullerene, because it's structure resembles the geodesic domes built by the famous architect and visionary Buckminster Fuller. Formally, this structure is called a truncated icosahedron, and consists of alternating hexagons and pentagons. Look at it closely and you will notice it looks more or less exactly like a soccer ball (or football to all but Americans – we have our own eccentric version of a football) with the same arrangement of pentagons and hexagons (Fig. 6).

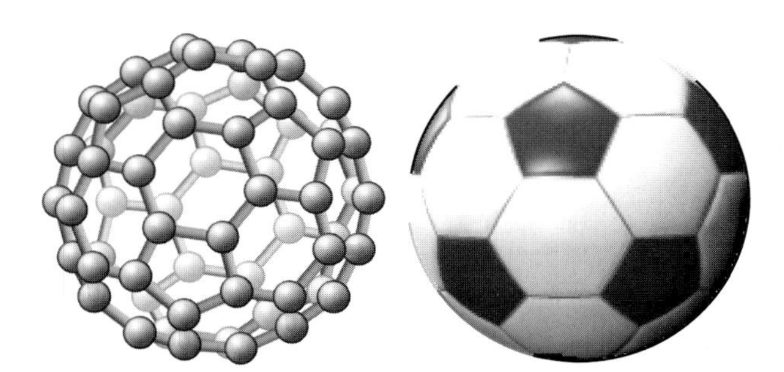

Figure 6 Chemical structure of Buckminsterfullerene-C60. Each carbon is bound to three other carbons in a pseudo-spherical arrangement consisting of alternating pentagonal and hexagonal rings, in the manner of a soccer ball. Hence its nickname, buckyball. C60 image reproduced courtesy of Samantha J. Shanley, University of Bristol.

Common to diamonds, graphite, carbon black and carbon nanotubes is the chemical formula – C_n – where n is the number of carbon atoms. All of the wildly different attributes of the various forms of carbon come about merely through the altered arrangement of those carbon atoms into molecules. Though carbon linked only to itself comes in a variety of forms, it hardly stops there: in combination with other elements, carbon forms about sixteen million different compounds. All life, as far as we know, is based on carbon chemistry.

If we can get this much utility out of carbon, how much more can we do with control over the placement of over all of the available elements?

An Alternative Nature

Biology, for all its genius, paints with a limited pallet – mainly carbon, hydrogen, nitrogen, oxygen and phosphorus, with some trace metals and salts thrown in for variety. These, in turn, are elaborated into only a few basic molecular types – proteins, nucleic acid, lipids, and carbohydrates. In contrast, chemists have the whole periodic table with which to work their magic. Until recently, their methods were relatively primitive and only small molecules could be efficiently manufactured. Now, nanotechnology seeks to unite chemists with physicist, engineers, and biologists to create molecular structures of unprecedented complexity and size. These structures can be used to create new materials and even nanoscale machines and artificial organisms. We are on the verge of creating what might be described as an alternate Nature.

All living things on this planet, from the tiniest virus to the tallest tree to the sperm whale in the ocean, share the same genetic code and substantially the same manufacturing scheme for putting together their various components. Out of the science of molecular biology came the recognition that there is substantial unity in the biochemical make-up of all creatures on the planet.

Now suppose that we could consciously control manufacture at the molecular level in the way that living things do. Inorganic components could be married with biomolecules. Building materials could have intrinsic self-repair capabilities. Skyscrapers could, in theory, be built such that the whole structure was covalently linked into one super molecule. Would this be a better way to build things?

Money Makes the World Go 'Round

The problem with some more enthusiastic blue-sky scenarios – the fly in the blue-sky – as always, is economics. In Neil Stephenson's sci-fi epic *The Diamond Age* [2], one of the first fiction works to focus on nanotechnology, buildings were grown from seed and raw materials with the help of molecular assemblers. Imagine, for instance, that you could grow a barn that way. Or you could hire a few Amish farmers and they will nail up a lovely barn for you in a weekend. Which really makes the most sense, from an economic standpoint? It would take a lot of barns to justify the development costs of the nanotech version.

Money, as the immoderate emcee in Bob Fosse's Cabaret reminded us, is what makes the world go 'round. Without its commercial appeal, nanotechnology would not go far. Nanopioneering products so far been modest in terms of products that have been produced and profits they have generated. Small nanoparticles are used to make sun-blocking cosmetics. Nanoparticles are also used as a slurry to polish silicon used in making semiconductors. Carbon nanotubes have been used as a reinforcing material in tennis rackets and in polyurethane. NanoTex stain-resistant fabrics are used to make clothing. Mercedes-Benz includes in its paint jobs nanometer-sized ceramic particles that makes the surface more scratch-resistant and helps keep it glossy. Similar particles are used in floor tiles. InMat had developed a thin coating for the inside of tennis balls that retards the loss of air pressure, extending their useful lifetime. There are potential applications of this process for everything inflatable, from car tires to helium balloons.

And what about self-cleaning windows? Talk about a boon to humanity! This invention relies on a coating, only 40 nm thick, which contains a photocatalyst that uses the sun's UV energy to break down organic debris that collects on the windows. A second feature of the coating is that it is chemically hydrophilic (water-loving). Water does not bead up on the glass, but sheets off evenly.

A scientist at the University of Queensland, Michael Harvey, has invented a nanoscale coating called Xerocoat that is actually a thin film of glass full of tiny bubbles. Xerocoat prevents fogging on such things as spectacles, automobile windows and bathroom mirrors. "We are taking nanotech out of the lab and putting it in the bathroom," says Harvey.

Nanotechnology has already established a foothold within your computer. The read-heads of newer hard drives are built by the nanoscale deposition of thin films of "giant magnetoresistant" material. This material has the property of changing its resistance to the flow of electricity when it encounters a magnetic field. The read-head glides over the hard-drive at speeds up to 80 miles an hour suspended on a cushion of air only 10 nm above the surface of the drive. The magnetically encoded data on the disk are translated into electrical current as the read-head flies along.

In terms of dollar volume, the most important nanotech products right now are probably nanoparticle catalysts used in the distillation of petroleum and its byproducts.

The real harvest of nanotechnology is yet to come. But technology does not develop in a vacuum. Ideas do not jump from the head of a scientist or engineer into reality. The translation of ideas to prototype to product requires great inputs of both toil and capital. And all of that ingenuity and investment may be wasted if the society or the market is not ready for the final product.

Who Knows About Nano?

"Everybody knows that nanotech is important," says Bob Gregg, executive vice president of FEI Corp., which makes electron microscopes. "Just mention the word, and you can get a meeting with anybody [in the federal government] in Washington D.C. Of course nobody knows what it means ..."

Despite a fair amount of media coverage, the promise of nanotechnology is not much appreciated by the general public. This was brought home to me last year, when I gave a presentation at a convention called 'Imaging and Imagining Nanoscience and Engineering', sponsored by the University of South Carolina in the city of Columbia. The night before my talk, as usual, I ran through my slides and gave a solitary performance for the benefit of my reflection in the window of my hotel room. This kind of concentration at night tends to get me too wired to sleep, so I went down to the bar. At that time, perhaps 25 people were assembled there in various states of intoxication. I quickly met up with a man who was staying at the hotel as a mentor for a convention of teen-age journalists. Despite being a journalist and therefore open to a wide variety of general information, this man claimed to have never heard the word "nanotechnology." Emboldened by a couple of beers, we proceeded to poll those assembled in the bar to determine if any of them understood the term. There was exactly one other patron there, other than myself, who admitted to knowledge of nanotechnology. An aerospace engineer, he opined that the university and state government were interested in nano only because they thought it would somehow provide jobs for South Carolina. This particular engineer was African-American; ironically, his female companion was at first very adamant that he not talk to us. Because she was a northern black recently moved to the South, she had the mistaken impression that my journalist compadre and I were engaged in some Southern whiteboy crusade to prove that black people were ignorant. I am quite sure that most people in the United States – white, black, Latino or indifferent – either have never heard of nanotechnology or have a vast misunderstanding about what it is about. I doubt that the rest of the world is any different.

Senator Ron Wyden (D.-Oregon), who is the co-author of the Twenty-First Century Nanotechnology R&D. Act, tells the story of one of his constituents, an elderly lady, who accosted him a local supermarket. "Senator Wyden, I don't know much about this 'nano-nology'," she says, "... but I'm glad you're doing it." Hopefully, this book will increase public knowledge about nanotech, the people behind it, and why they're doing what they're doing.

Nanotechnology requires not only scientists and engineers, but also entrepreneurs with vision, not to mention patent lawyers and marketing agents. Right now, nanotechnology is the sphere of a small number of entrepreneurial companies and a few large giants, like IBM, that have an eye for the future. An economic depression, a World War or an overwhelming natural disaster, like global warming, have the potential to derail the technological future in the making. At least for a while.

The Promise of Nano

Warnings in place, let us examine some of the claims that are made for nanotechnology in the near future and beyond.

The promises of nanotechnology are ubiquitous in nature: To make that point, Table 2 lists the use of "nano" as a prefix in words that are often used in the nanotech domain, even if they haven't yet quite made it into *Webster's Dictionary*. All of the terms below were actually abstracted from this book. Like any good writer, I am not averse to an occasional neologism if I can't find an extant English word that seems to work just right. However, I do not claim any of the words below as my own.

Table 2 The proliferation of "Nano" as a Prefix.

nanoage	**nanocrystals**	**nanomagnetic**	**nanoscale**
nanoarray	nanocube	nanomanipulator	nanoscience
nanoassembly	nanodevice	nanomaterial	nanoscope
nanobacteria	nanodivide	nanomedicine	nanosecond
nanobiologist	nanodomain	nanometer	nanoshell
nanobiomedicine	nanoelectromechanical	nanomicelle	nanostructured
nanobiotechnology	nanoelectronics	nanoparticle	nanostructures
nanobot	nanoencapsulation	nanoparticulate	nanoswarm
nanocapsule	nanofabrication	nanophase	nanosystems
nanocassette	nanofibers	nanoplatelates	nanotechnology
nanocatalyst	nanofilter	nanoporous	nanotool
nanocomponent	nanofluidics	nanopowder	nanotube
nanocomposite	nanolayer	nanoproduct	nanotweezers
nanoconnections	nanoliter	nanoreactor	nanowire
nanocosm	nanolithography	nanoreplicator	nanoworks
nanocrystalline	nanomachine	nanorobotics	nanoworld

Table 2 is hardly an exhaustive list, particularly if you start including the names of companies – NanoInk, NanoSphere, Nano-Opto, Nanoproprietary, Nanoset, Nanosys, etc. – or the names of products – Nano-fur, NanoReader, NanoSolve, Nanobac.

„Micro-", as a prefix – as in microscope or microbe or microelectronics – has been part of the language for many years. "Micro" actually has a technical meaning – it means one-millionth. A micron, for instance, is one millionth of a meter. In popular usage, however, "micro-" has devolved into a prefix meaning simply "very small." Even in technical usage, this is true.

"Nano-" also has a technical meaning – it means one-billionth. Since nanoscale engineering has become possible, "nano" is undergoing a linguistic expansion that is overtaking micro. This transition is being accelerated quickly by firms and marketing trying to take advantage of the buzz surrounding nanotechnology. Nano-, in popular usage, will perhaps in time come to mean very, very small, but not necessarily exactly nanoscale. At the same time, micro- remains a part of the language. Thus, we talk about atomic force microscopes, even though they are used primarily to image objects – atoms and molecules – that are measured in nanometers and even angstroms (one-tenth of a nanometer). Likewise, the term "microfabrication" is often used, even when the subject is really nanofabrication. It is unlikely that this confusion in the language will be resolved anytime soon.

As almost every technology will soon have some nano-component, the term "nanotechnology" may ironically become obsolete, as the word will seem to contain an internal redundancy. However, a residue of nano-prefixed words will be left in the language forever.

Besides basic materials, nanotechnology already encompasses medicine, electronics, energy production, and computing.

Nanoparticles already under development deliver drugs in a targeted fashion to specific cells in the body. Thus, it may be possible to kill cancer cells with a potent toxin without significant damage to normal cells. Nanoscale devices will eventually be employed as drugs or for drug delivery; in assays used for medical diagnosis, drug discovery, or basic biological research; as contrast agents for MRI imaging; and in imaging instruments, like X-ray devices. Exquisitely sensitive biosensors will allow the monitoring of a thousand different parameters of our health from a single drop of blood. Similar sensors, cheap and ubiquitous, will monitor the environment for dangerous chemicals, toxins, and even viruses – serving as a kind of external immune system. Cameras with nanoscale components will take exquisitely detailed pictures of our tissues from within our bodies.

Genetic sequencing will become extremely rapid and cheap. It may be possible to have your own personal genome sequenced over the weekend for about the cost of many current medical tests. With that information, your doctor will tell you probably more than you wished to know about your vulnerability to cancer, diabetes, and degenerative diseases.

Electronics will become molecular, with devices connected by nanoscale wire. Instead of being elaborately manufactured, electronic elements may self-assemble, or be printed onto flexible sheets. The desktop computer may be enclosed into a thin surface laminated right onto the desktop, taking up no room whatsoever. Electronics may be built into your clothing.

New solar energy devices that mimic photosynthetic pathways will become available, reducing our reliance on fossil fuels. Nano-enabled solar cells will extract hydrogen from water to be used as in fuels cells. Other energy devices will convert ambient heat into electricity. Already, a biothermal battery for medical implants is under development that recharges itself by converting body heat to energy. Other medical implants may run off of the body's own biochemistry, requiring no other power source.

As electronic circuitry becomes molecular, it will become possible for our nervous system to exchange information with electronic devices. The first beneficiaries of this technology may be paraplegics and quadriplegics who will be able to use electronics to circumvent their damaged spinal cords. A device in the brain will be able to determine the person's intentions and convert these into electronic signals to the arm and leg muscles, allowing these once helplessly paralyzed people to resume more less normal functioning.

As the technology improves, we may avail ourselves of memory implants that help us to remember information. Today, many people could not function effectively in their jobs or personal lives without the aid of computers. Tomorrow, this cooperation may be furthered by an actual physical or electronic connection between the brain and the computer.

Already well developed is a nanotech field called "spintronics," in which information is conveyed by the spin of electrons, rather than the charge (see Chapter 7). Further along, we may have quantum computing, where processing is actually performed through the interaction of quantum states within atoms. This would allow the compression of computing power on a truly astonishing scale. Computing power equivalent to all of the computers ever manufactured, including the human kind, might be represented in a few cubic centimeters of matter.

With the remarkable advances promised by nanotechnology, it is not surprising that governments around the world want a piece of it. In 2003, U.S. President Bush signed into law the 21st Century Nanotechnology Research and Development Act, which promised $3.7 billion in federal funds for nanotech programs. About $ one billion of this will be spent in fiscal year 2005. The Japanese government is matching U.S. funding dollar for dollar (or rather, the yen equivalent). The Chinese and Koreans have very active nanotechnology development programs. Europe, with some trepidation, is following the path of its trading partners; overall public funding is reportedly on the order of 700 million annually for nanotech

Skeptics

The Europeans have a recent history of being technology skeptics. For example, environmentalists on the continent reviled genetically modified grains as "Frankenfoods," on the theory that modified genes in maize or soybeans would have potentially disastrous health effects on the consumer. The various Green parties were successful in at least delaying the introduction of these crops into Europe as well as the sale of genetically modified produce in the grocery stores. So far, the fear of adverse health effects has proven unfounded. Also, some of the benefits of genetically modified food have been hard to deny; for instance, rice modified to produce higher levels of vitamin A should combat the blindness related to deficiency of this vitamin in the Third World.

No less a luminary than Charles, Prince of Wales, has raised alarms about the potentially disastrous effects of nanotechnology in the environment. Environmental groups, notably one called ETC (which is based in Canada, a European country located by chance in North America), have actually called for a ban on the further development of nanotechnology, claiming environmental and social effects of the technology have not been adequately considered. "It is important for this rapidly evolving technology to identify and resolve safety concerns (real or perceived) at the earliest possible stage. Successful exploitation of nanotechnologies needs a sound scientific basis for both consumer and commercial confidence," says the Commission of European Communities [3]. In Chapter 11, called Fear of Nano, we will consider some of the potential dangers of nanotechnology and ways to ameliorate them.

Contemporaneous History

Writing a book about a rapidly expanding technology is always an exercise in frustration, as it is sure to be out of date as soon as it is published. Nanotechnology is particularly difficult because it is so all-encompassing that virtually every industry will eventually be affected. Try as I might to give a comprehensive view, some areas will be left out. I will be satisfied, however, with a sort of contemporaneous history, a journalistic report on events as they are now occurring.

Science and technology are not realized in a serene, ivory tower environment; rather, they are the products of intellect supplied with real world resources. Competition for those resources, whether from government, academic or private sources, is as fierce as anything found in Darwin's wild world. A pioneer of any kind is generally blessed with an active curiosity and a sense of adventure; naked aggression is also a valuable trait. This is a story about politics and personality as much as it is about science and engineering.

I would like to end this chapter with an apology to all of the Nanotech Pioneers whose names don't appear in this book. Obviously, a book of finite length cannot mention everybody whose work is interesting; choices have to be made. Chance and circumstance also play a part in what went in and what had to be left out.

References

1 *The Journals of Lewis and Clark*, edited by Frank Bergen, Penguin Books, 1989.

2 Stephenson, Neil, *The Diamond Age*: or *A Young Girl's Illustrated Primer*, Bantam Books, 1995.

3 *Towards a European strategy for nanotechnology*, A Communication from the Commission of European Communities, 12.5.2004. ftp://ftp.cordis.lu/pub/nanotechnology/docs/nano_com_en.pdf

Chapter 2
The Visionaries

Nanotechnology has its origins in many fields – in the realm of quantum physics, in biotechnology, in industrial work on branching polymers and combinatorial chemistry, in the developmental of powerful electron microscopes, and later scanning microscopes. None of this work was called nanotechnology at first, and the connections between these various disciplines were not apparent. One of the first applications of nanotechnology was the fabulous stained glass fabricated in the Middle Ages, which made use of small gold nanoparticles to create luminous red pigments, so it is a fact that nanotechnology has been around for centuries. But the possibilities inherent in the field have only been appreciated recently.

Richard Feynman

There were people who saw into the future. The first of these was the quizzical free-spirited physicist named Richard Feynman. He won a Nobel Prize for his work in quantum electrodynamics, although at one time he was judged mentally deficient by a U.S. Army psychiatrist. He is probably known best to the American

Figure 7 Richard P. Feynman. Image reproduced courtesy of the Nobel Foundation.

The Nanotech Pioneers. Steven A. Edwards

ISBN: 3-527-31290-0

public as the man at the inquiry into the Challenger disaster who showed that the malfunction of an O-ring, a fifty-cent part, cost seven people their lives plus the destruction of space shuttle. By dipping an O-ring into ice water, Feynman demonstrated that O-rings lose elasticity at low temperatures like those on launch day, and therefore lose their sealing function. Richard Feynman was that rare commodity, a scientific genius with common sense.

Feynman was also known to have a great sense of humor (his autobiography is titled, "*Surely You're Joking, Dr. Feynman* [1]). Which is probably why most of his audience at the American Physical Society on December 29th, 1959, did not take his lecture that day too seriously. The lecture was entitled, "There's Plenty of Room at the Bottom [2]."

"Why...," asked Feynman in his lecture, "... can we not write the entire 24 volume of the *Encyclopedia Britannica* on the head of a pin ?" [2]. He proceeded to calculate that there was plenty of room on the pin, if only we had the means to do so.

"If you magnify it by 25 000 diameters ...", said Feynman, "... the area of the head of the pin is then equal to the area of all the pages of the *Encyclopedia Britannica*. Therefore, all it is necessary to do is to reduce in size all the writing in the Encyclopedia by 25 000 times. Is that possible? The resolving power of the eye is about 1/120 of an inch – that is roughly the diameter of one of the little dots on the fine half-tone reproductions in the Encyclopedia. This, when you demagnify it by 25 000 times, is still 80 angstroms in diameter – 32 atoms across, in an ordinary metal. In other words, one of those dots still would contain in its area 1000 atoms. So, each dot can easily be adjusted in size as required by the photoengraving, and there is no question that there is enough room on the head of a pin to put all of the *Encyclopedia Britannica*."

Got that? Feyman, like Einstein, was a brave intellect, willing to follow the truth wherever it led. In fact, he offered a $1000 prize to the first person who could reduce the written word by 25 000 times, sufficient to put the *Encyclopaedia Britannica* on the head of a pen. The challenge was eventually met and the prize money collected. The device used to do the writing was an electron beam, of the kind used to make masks for semiconductor chips. Figure 8 demonstrates another means to do the deed, nanolithography. Nanolithography essentially employs the an atomic force microscope tip, which looks disturbingly like a fountain pen, except that its tip is carved down such that it is only contains a few atoms at its sharpest point.

These days, we are familiar with the idea of information, if not actual writing, that fits into tiny spaces. Just think how many gigabytes will fit into a current hard-drive. But in 1959, the few computers that existed filled entire rooms, and required dedicated air-conditioning systems to keep them from overheating. Feynman's ideas were, at that time, quite radical.

In his short essay, Feynman anticipated much of what we see developing in nanotechnology today. He pointed out that the electron microscopes of the day were operating well below their limit of resolution. He challenged physicists to improve the resolution such that the machines would be powerful enough that

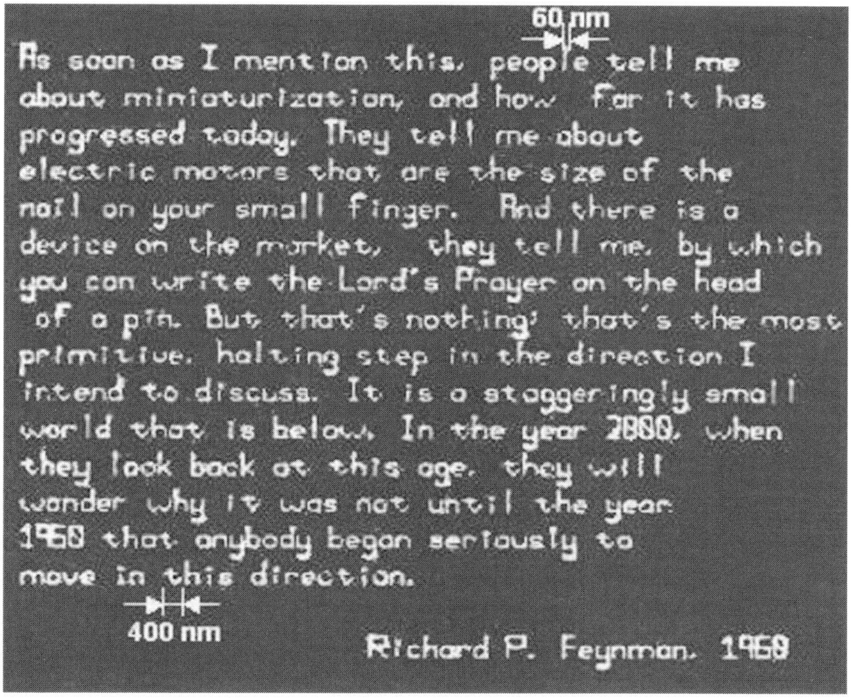

Figure 8 Part of Richard Feyman's Lecture etched on the head of a pin using dip pen nanolithography. Courtesy of NanoInk, Inc.

biologists could see directly the interaction of molecules within cells, that DNA might be sequenced simply by looking at it. He even suggested that miniature robots that could operate within our bodies – an idea that Isaac Aasimov later elaborated in his science fiction vision, "*The Fantastic Voyage*."

Feynman suggested a top-down approach toward manufacture whereby miniature tool sets would be used to make more miniature tool sets, which would in turn make yet smaller tools, until finally we would be able to work at the nanoscale. He saw nothing in the principles of physics that would prevent us from directly manipulating atoms. Forty years later, Don Eigler used a scanning tunneling microscope tip to arrange xenon atoms on a nickel crystal to spell out the letters I.B.M., thereby demonstrating in practice what Feynman had predicted.

"It is a staggeringly small world that is below," said Feynman. "In the year 2000, when they look back at this age, they will wonder why it was not until the year 1960 that anybody began seriously to move in this direction [2]."

Actually, nobody seriously began moving in that direction for another twenty years. Feynman's lecture, though published in 1960, was largely forgotten. Feynman's own autobiography doesn't even mention the *Room at the Bottom* talk – it was just one of many sparks sent flying into the world by Feynman's prodigious intellect. The world was not quite ready for nanotechnology *per se*. But related fields were pushing on.

K. Eric Drexler

Feynman's talk was eventually re-discovered by an MIT graduate student named K. Eric Drexler, although not until he had come to nanotechnology by a completely different direction.

When Feynman spoke in 1959, it had been only six years since James Watson and Francis Crick had proposed a double helix as the structure of DNA [3]. Molecular biology was in its infancy. The 1950s and early 1960s saw an explosion of information, with the general scheme called the Central Dogma, in which genetic information was embedded in the sequence of DNA, copied into messenger RNA, and translated into the sequence of proteins.

Nature provided biologists with an indispensable tool in the form of restriction enzymes – bacterial proteins that could cut DNA in a sequence-specific manner. This allowed researchers to cut and splice DNA in a manner similar to editing a film. Particular stretches of DNA could be amplified by splicing them into bacterial plasmids – small, circular stretches of DNA carried by bacteria that usually carried antibiotic resistance genes. By various manipulations, bacteria could be induced to replicate plasmids on the order of thousands of copies per cell.

Recognizing the commercial potential of the ability to rearrange DNA, Stanley Cohen and Herb Boyer of Stanford University, applied for the first patent in "recombinant DNA" technology in 1974. By 1976, Boyer and Robert Swanson had founded the first biotechnology company, Genentech, with the help of venture capitalists. By 2003, Genentech had annual sales of $2.8 billion dollars, with protein drug products to treat cancer, heart disease, diabetes, and asthma.

Molecular biology greatly stimulated the imagination of K. Eric Drexler, who was not a biologist, however, but an engineer. He was to write the first journal article [4] discussing "nanotechnology" (although the word was not actually used). His article, entitled 'Molecular Engineering: An approach to the development of general capabilities for molecular manipulation,' was published in 1981 in the prestigious journal *Proceedings of the National Academy of Sciences.* To Drexler, life

Figure 9 K. Eric Drexler. Image reproduced courtesy of the Foresight Foundation.

was the existence proof that machines could be designed and built atom-by-atom. The ribosome, for instance, is a general-purpose protein factory in which the blueprint inferred by the messenger RNA is translated into reality. Drexler made direct comparisons between macroscopic machine parts and the components of biological cells, as shown in Table 3, which is reproduced from his paper.

Table 3 Comparison of macroscopic and microscopic components.

Technology	**Function**	**Molecular example(s)**
Struts, beams, casings	Transmit force, hold positions	Microtubules, cellulose, mineral structures
Cables	Transmit tension	Collagen
Fasteners, glue	Connect parts	Intermolecular forces
Solenoids, actuators	Move things	Conformation-changing proteins, actin/myosin
Motors	Turn shafts	Flagellar motor
Drive shafts	Transmit torque	Bacterial flagella
Bearings	Support moving parts	Sigma bonds
Containers	Hold fluids	Vesicles
Pipes	Carry fluids	Various tubular structures
Pumps	Move fluids	Flagella, membrane proteins
Conveyor belts	Move components	RNA moved by fixed ribosome (partial analog)
Clamps	Hold workpieces	Enzymatic binding sites
Tools	Modify workpieces	Metallic complexes, functional groups
Production lines	Construct devices	Enzyme systems, ribosomes
Numerical control systems	Store and read programs	Genetic system

Source: K. E. Drexler, 'Molecular engineering: An approach to the development of general capabilities for molecular manipulation.' *Proc. Natl. Acad. Sci. USA* 78: 5275–5258 (**1981**).

The nanomachines within a human cell, though they have their counterpart in the macroscopic world, operate in a different environment. Our physical intuition with regard to forces like gravity, friction and inertia are of little use in considering these nanodevices. Because they are floating in a liquid medium and because they are very small, gravity has little impact and friction has little meaning. Viscosity is much more important. Nanomachines in the cell are built with atomic-scale features and exist in an environment in which collisions occur constantly with other atomic scale-objects. Nothing stays still; molecules are vibrating rapidly at all times. Surface effects predominate at the nanoscale, since every atom is at most a few atoms away from the surface.

In 1986, Drexler published *The Engines of Creation: the Coming Era of Nanotechnology* [5], in which he explained his ideas about nanotechnology in a way that was accessible to the public (a more technical work, *Nanosystems: Molecular Machinery, Manufacturing, and Computation*, was published by Drexler in 1992). This was the first real attempt to define nanotechnology as a field unto itself. The book quickly became popular among both graduate students and science fiction fans (who are frequently the same people). Drexler argued that nanotechnology had the potential to usher into the world an unprecedented era of abundance and longevity.

Central to Drexler's argument was the idea of a Universal Assembler, a hypothetical device that would enable molecular manufacturing by snapping together molecules in the manner of Lego blocks. The Assembler was not entirely a Drexlerian innovation; like so many novel ideas, the Assembler had its origin in the mind of mathematician and game theorist John Von Neumann. Von Neumann's assembler consisted of two central elements: a universal computer and a universal constructor. The computer directed the behavior of the constructor which, in turn, was used to manufacture both another universal computer and another universal constructor. The program code contained in the original universal computer would then be copied to the new universal computer and executed.

The ideal of an assembler that can build itself is already being realized, in part, at the macroscopic scale. "The potential impact of intelligent machines is magnified by that fact that technology has reached the point where intelligent machines have begun to exhibit a capacity for self-reproduction," notes James S. Albus, National Institute of Standards and Technology (quoted in Ref. [7]). Albus points out that computers are now instrumental in designing, testing and manufacturing other computers, as well as other machines. IBM has built a prototype computer assembly plant where the only mechanical input from human beings is on the loading dock

Drexler's contribution was to scale the assembler, in theory at least, down to molecular size. Drexler took very seriously Feynman's point that nothing stood in the way, finally, of moving atoms about one at a time, and building materials with atomic precision. After all, is this not what biology is all about? The molecular assembler must be able to manipulate structures with atomic precision. The universal molecular constructor has both a positional capability and some sort of dynamic chemistry that makes it able to manipulate individual atoms. In other words, an arm and a hand, with very tiny fingers. Another necessary attribute would be some means of programming the assembler. It would also be desirable to have some sort of macroscopic interface with people or computers whereby the assembler could respond to given commands or communicate its needs.

A single such assembler working night and day would not have much effect, however, any more than a single ribosome supplied with an endless supply of mRNA and amino acids could add a significant amount of protein to the world.

In order to be useful, the first assignment for the assembler would be to make more copies of itself. Each new assembler, in turn, could replicate and the copies could replicate until very rapidly there would be enough of the new devices to enable a revolution in manufacturing.

Nanotechnology is often separated into "wet" nanotechnology, which is equivalent to biotechnology, and "dry" nanotechnology – building nanoscale devices without the aid of biomolecules or cells. Drexler referred to the latter as second-generation nanotech in *Engines of Creation* [5]. He imagined that dry nanotech would involve "mechanosynthesis" – literally building machines from the bottom up using assemblers that would place one atom (or molecule) after the next to the other in a predetermined order.

Drexler and followers seem to have fixated on "diamondoid" mechanosynthesis for molecular manufacturing, a kind of directed, dry chemistry in which carbon-based machines are created atom-by-atom. When questioned, however, Drexler himself admitted that the first assemblers might involve something re-engineered from living cells.

"There seems to be a confusing thought that there is only one unitary pathway to artificial molecular machine systems ...," he told me in an e-mail interview, "This would be incorrect. There are multiple paths, many of which build on one another. My first technical publication in this area, my 1981 article for the Proceeding of the National Academy of Sciences was indeed based on a biological model. You are correct that one potential pathway is based on synthetic ribosome-like and enzyme-like complexes. In fact, this is an area where some of the most interesting research is being conducted today."

Ralph Merkle

One of Drexler's early converts was Ralph Merkle, by all accounts a brilliant man who seems to be living a little ahead of his time. As a graduate student at Stanford, he was credited with being the co-inventor of public key encryption, a method of computer security that has become widely used to protect the contents of e-

Figure 10 Ralph Merkle with molecular models. Image used courtesy of Ralph Merkle.

mail and other computer documents. But Merkle received his doctorate in 1979, before there was an Internet.

Merkle worked at Xerox PARC (Palo Alto Research Center), a legendary laboratory that was responsible for many of the innovations now used in computing, including the mouse, the graphical interface, and the laser printer. Few of these inventions did much for Xerox's bottom line, it should be noted; Apple Computer did more with Parc's innovations than Xerox did.

Merkle continued to work on computer security at Xerox, but was also able to spend much of his time on nanotechnology. Although nanotech had no particular place in the strategic thinking of Xerox, Merkle's research, as he puts it, "... was just part of the ambience of work in a major research center." This sort of tolerance is becoming rare at large companies in the 21st century.

Drexler, Merkle and others, spent a lot of time creating designs of molecular-scale gears and rotors and similar machine parts that could be built as soon as the universal assembler was available (Fig. 11). The assembler, it was thought, would lead to a "two-week revolution" in which an amazing diversity of nanoscale devices would suddenly become enabled – similar to the Cambrian Explosion, when the body-plans of living creatures expanded into their myriad forms.

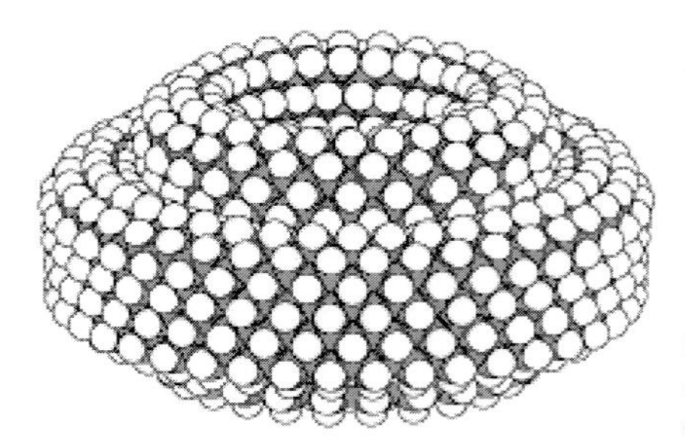

Figure 11 Hydrocarbon bearing. Each circle represents an individual atom. Image reproduced courtesy of Ralph Merkle.

Drexler was enamored of the logarithmic growth phase exhibited by bacteria. Given a continuous supply of nutrients, *E. coli*, for example, will double its population about every 20 minutes or so. In theory, *E. coli* could cover the earth with a continuous lawn of bacteria within a matter of days. Assemblers could replicate in a similar fashion, it was thought. However, such a scenario is not without peril.

To his credit, Drexler warned of the possibilities inherent in nanotechnology, even though he is an ardent booster. In *Engines of Creation* [5] he pointed out some clearly dystopian possibilities inherent in nanotechnology:

> "Plants" with 'leaves' no more efficient than today's solar cells could outcompete real plants, crowding the biosphere with an inedible foliage. Tough omnivorous 'bacteria' could out-compete real bacteria: They could spread like blowing pollen, replicate swiftly, and reduce the biosphere to dust in a matter of days."

The negative consequences of nanotechnology alluded to by Drexler had a powerful effect on the minds of science fiction writers, who took the ball and ran with it, culminating in 2002 in the novel *Prey* [8], written by best-selling author Michael Crichton. *Prey* features a swarm of nanocreatures with a distributed intelligence and a nasty predilection for disassembling human beings. Many in the nanotech community became concerned after a writer with Crichton's following picked up the theme of runaway replicators. Vicki Colvin, executive director of the Center for Biological & Environmental Nanotechnology at Rice University, said in testimony before Congress that *Prey* illustrates "... a reaction that could bring the growing nanotechnology industry to its knees: fear. The perception that nanotechnology will cause environmental devastation or human disease could itself turn the dream of a trillion-dollar industry into a nightmare of public backlash." (quoted in [7]).

After the publication of *Engines of Creation* [5], Drexler tried to undo some of the negative publicity he had created. In an Afterword added in 1990, he says, "Certain scenarios and proposals in the last third of *Engines* could bear rephrasing, but at least one problem is presented misleadingly. Page 173 speaks of the necessity of avoiding runaway accidents with replicating assemblers; today I would emphasize that there is little incentive to build a replicator even *resembling* one that can survive in nature." The damage, however, had been done.

Today, Drexler goes farther, insisting that self-replication of assemblers is not even necessary to fulfill his vision of nanotechnology. "... it turns out that developing manufacturing systems that use tiny, self-replicating machines would be needlessly inefficient and complicated. The simpler, more efficient, and more obviously safe approach is to make nanoscale tools and put them together in factories big enough to make what you want."

Whether or not Drexler's nanobots assemblers are ever built, there is no doubt that he had an enormous impact in presenting the possibilities inherent in nanotechnology to the public and in making the field sexy to a generation of researchers. William Illsey Atkinson, in his book *Nanocosm* [9], throws a backwards compliment to Drexler and his followers, "By piquing interest, first in the broader pubic and then in mainstream scientists, the [nano]boosters have advanced basic nanoscience and accelerated the commercialization of its discoveries. The boosters, bless their goofy hearts, have thrown open the doors to more disciplined imagination. In so doing, they have, (however briefly) filled a real need."

Many in the nanotech community are not enthusiastic about Drexler, or his ideas. In part, this was due to the baggage that they come with, the possibility of runaway replicators, and the science fiction aura with which *Engines of Creations* was presented. Moreover, Drexlerian nanotechnology has become popular with the Extropians, a California transhumanist technology cult which espouses a lot of unlikely technology fixes to lengthen life, such as freezing one's dead body in hopes of resurrection, or uploading back-up personality into a computer, in case your physical body had a misfortune. In *Engines of Creation* [5], Drexler had promised nanotech-enabled cell repair machines. "People who survive intact until the time of cell repair machines will have the opportunity to regain youthful health and to keep almost as long as they please. Nothing can make a person (or anything else) last forever, but barring severe accidents, those who wish to do so will live a long, long time." Drexler and the Extropians, led by Maximum More, made a perfect match. This kind of breathless acceptance of sci-fi scenarios, however, does not play well with either academic scientists, entrepreneurial businessmen, or importantly, venture capitalists.

Ralph Merkle, Drexler's associate, is a director of the cryogenics firm Alcor, which freezes people – no, excuse me, *vitrifies* – corpses in the hope that they can eventually be rejuvenated with advanced technology, including nanotech cell repair machines. As Merkle points out, your probability of dying (permanently) if you don't undergo this procedure is 100 %, whereas there is a finite probability that the technical problems that led to your terminal disease may be correctable at some point in the future, and that your body can be reanimated using a procedure that has not yet been invented. Many people, in fact, elected the simpler, cheaper alternative of simply having their heads frozen, the idea being that the remaining tissue could be regenerated using stem cell technologies. Or perhaps, the personality could simply be uploaded into a computer, a process envisioned by robotics engineer Hans Moravec [11] and others.

Ray Kurzweil

Ray Kurzweil has had a long career as an inventor, audio engineer, and technologist. One of his more useful inventions was a reading machine for the blind – a machine that scans written materials and then reads it out in a synthesized voice. Supposedly in response to a challenge from blind singer/songwriter Stevie Wonder, Kurzweil went on to develop the first ROM sampling keyboards, marketed by Kurzweil Music systems. Kurzweil is probably best known to the public, though, for a series of futurist books he has written, including *The Age of Intelligent Machines, The Age of Spiritual Machines,* and the still-to-be released book, *The Singularity is Near.* Kurzweil's website. Kurzweil's website, www.kurzweilai.net, is a compendium of information on all things weird and wonderful, technology-wise, with a particular focus on artificial intelligence. On the lecture circuit, Kurzweil is

Figure 12 Ray Kurzweil. Courtesy of Ray Kurzweil and Kurzweil Technologies.

often paired with Bill Joy (see below), the alumnus of Sun Microsystems who warns of the dangers of advancing technology, while Kurzweil sounds the upbeat, techno-salvation counter-argument.

Kurzweil is one of the latest in an ongoing series of eccentric American inventors that would include such luminaries as Tom Edison, Nikola Tesla, Howard Hughes, and Buckminster Fuller. On the other hand, with respect to nanotechnology, it would be a mistake to conclude that he has actually contributed anything other than his opinions. Likewise, it would be a mistake to discount those opinions, because Kurzweil is a well-regarded and obviously intelligent and informed observer, as well as being highly visible.

Ray Kurzweil has taken up Drexler's vision that nanotech has the potential to offer effective immortality. Along with physician Terry Grossman, he has written a book called *Fantastic Voyage: Live Long Enough to Remember Forever* [9], which offers baby boomers advice on how to extend their life long enough to make it into the nanotech age of medicine. Diet and exercise, they maintain, is the First Bridge to this new world of extended lifelines. Biotechnology, especially stem cell therapies, would be the Second Bridge, while nanomedicine would be the Third Bridge.

Kurzweil is also known for his "Law of Accelerating Returns" [10]. He argues that technological progress proceeds much faster than we intuitively view it to happen. He argues that evolutionary processes, whether biological or technological, progress at exponential rates. In fact, he points out that technology is the ultimate result of evolution, since we, the technology-creating species, are a result of evolution. The returns on exponential progress (speed or cost-effectiveness) are likewise exponentially increasing. Therefore, instead of 100 years of progress during the 21st century, we will achieve something like 20 000 years of progress, measured in arbitrary units that seemed intuitive in the 2000s.

Different fields cross-fertilize each other; biotechnology has been fed by computer science, for instance, and now biotechnologists are trying to make computers with DNA. The information handling necessary for the human genome project would never have been possible without the simultaneous advances in the availability of computer power. On the other hand, knowledge about the brain has fed into computer science through the development of "neural network" software programs that exhibit learning capabilities.

Kurzweil argues for the coming of the Singularity, an event apparently predicted by John Von Neumann, the great mathematician and game theorist. The Singularity would consist of technological progress so rapid that it would be beyond the capability of humans to control or predict.

Vernor Vinge, a mathematician and science fiction author, was the first to popularize the idea of such a technological Singularity. For Vinge, the critical catalyst locking in the Singularity is the creation of greater than human intelligence through the cooperation (or even the merging) of man and machine. Kurzweil sees the Singularity as a more complicated culmination of many different technologies feeding into each other, including computer science, biotechnology, medicine, and nanotechnology. While Kurzweil is an optimist, Vinge is more cautious.

When asked if he would pull the plug on greater than human intelligence though, Vinge said, "No. There would be some researchers who would pursue the goal, so the global answer is, 'no'." In other words, the Singularity is inevitable [11]. Similar statements have been made about nanotechnology – since the means to develop this technology is already available around the world, it would not be in any nation's interest to choose not to develop it. Global regulation is, therefore, seen as impossible.

One of the dangers that we visit in Chapter 11, Fear of Nano, is the possibility that nanotechnology will accelerate the arrival of the Singularity.

Criticism of the Drexlerian Vision

Not everybody is sanguine about the idea accelerating technology. Bill Joy, a founder and former chief technology officer for Sun Computers, wrote a now famous article for *Wired* magazine, titled "Why the Future Doesn't Need Us," in which he detailed his fears about further progress. "The 21st-century technologies – genetics, nanotechnology, and robotics (GNR) – are so powerful", said Joy, "that they can spawn whole new classes of accidents and abuses. Most dangerously, for the first time, these accidents and abuses are widely within the reach of individuals or small groups. They will not require large facilities or rare raw materials. Knowledge alone will enable the use of them [12]."

Specifically, about Drexler's vision of nanotechnology, Joy said:

> " ... Rereading Drexler's work after more than 10 years, I was dismayed to realize how little I had remembered of its lengthy section called 'Dangers and Hopes,' including a discussion of how nanotechnologies can become 'engines of destruction.' Indeed, in my rereading of this cautionary material today, I am struck by how naive some of Drexler's safeguard proposals seem, and how much greater I judge the dangers to be now than even he seemed to then."

"How soon will we see the nanoscale robots envisioned by K. Eric Drexler and other molecular nanotechnologists?" asked Nobel-laureate Richard Smalley rhetorically [13] in a *Scientific American* article. The answer, he maintains is never. He argues that there are two fundamental problems, which he refers to as "fat fingers" and "sticky-fingers." Smalley says that robot will not be able to manipulate molecules within molecular spaces, because grippers cannot be built of the dimensions required, given that they must also be constructed of molecules. He also argues that a nanobots capable of attaching its fingers to a particular molecule, presumably through some sort of chemistry, would not necessarily be able to let go, making assembly impossible. What is needed in order to make a universal assembler is really "magic fingers" according to Smalley.

In an open letter to Eric Drexler published in *Chemical and Engineering News,* Smalley went further, "You and people around you have scared our children. I don't expect you to stop, but I hope others in the chemical community will join with me in turning on the light, and showing our children that, while our future in the real world will be challenging and there are real risks, there will be no such monster as the self-replicating mechanical nanobot of your dreams."

Whether or not Smalley's criticisms are valid (and I, for one, would not argue chemistry with a Nobel-prize winning chemist), it is worth asking why Smalley would spend his valuable time arguing the point. Smalley won his Nobel prize for the discovery of buckminsterfullerene, the soccer ball-shaped carbon molecule that has become an icon of nanotechnology. Smalley has admitted publicly that *Engines of Creation* stimulated his own interest in nanotechnology. He has even contributed talks to conferences sponsored by the Foresight Institute, a public interest group founded by Drexler and Christine Peterson, to serve as an information source for the field of molecular nanotechnology. Ralph Merkle suggested to me that Smalley, as a very prominent nanotechnologist who has taken on public policy roles, has a vital interest in disentangling the field from the negative possibilities implicit in the Drexlerian vision of nanotechnology. Jim Von Ehr, CEO of nanotech company Zyvex, observed more graphically that "If a mosquito buzzes around and around your head long enough, you get the urge to swat it."

Eric Drexler has not been willing to be tied to any single company hoping to commercialize nanotechnology, apparently preferring to maintain his role as academic visionary, except that he is not attached to any academic institution either. Critics suggest that Drexler is not willing to do the hard work required to turn his dream into reality.

Drexler was, however, directly instrumental in the Zyvex Corporation, which was founded by Von Ehr for the express purpose of making the molecular assembler a reality. Von Ehr heard one of Drexler's lectures in 1993. Not impressed with the sci-fi razzle-dazzle of the lecture, he asked Drexler for a more technical treatment of the subject as was referred to *Nanosystems* [6] – Drexler's weighty tome that many refer to, but few have actually read. Von Ehr is one of those few.

James Von Ehr

James Von Ehr is the sort of down-to-earth multimillionaire with whom you could actually be comfortable using his first name. And so we will. Jim made more money than he had ever dreamed possible when he sold his software company Altsys to Macromedia in 1995. Rich, at the age of forty-five, he felt it was way too early to retire. He thought it would be an opportune time to start a nanotech company. Had Jim been a physicist instead of a software developer, it is unlikely that he would have considered a molecular assembler as an achievable goal for an entrepreneurial company. But at that point, he was blissfully naïve.

Of course if cloning pioneer Ian Willmut had been a developmental biologist instead of a vet, he would have known that cloning was impossible because cell differentiation is irreversible. But he didn't know that, and so Dolly – the first cloned animal – was born.

Jim interviewed a number of prominent university professors to join his new nanotech venture but was surprised to find none that was interested. He even

Figure 13 Jim Von Ehr. Courtesy of Zyvex Corp.

offered Eric Drexler a chance to help physically realize his vision of nanotech, but Drexler also declined. Says Drexler (who always sounds like the college professor he has never been) "There have always been two distinct, but overlapping, areas of my work. One is the theoretical technical research as demonstrated in the *PNAS* paper, other journal papers, and the technical text *Nanosystems*. The second area is to heighten public understanding and encourage discussion to help society prepare for a future of disruptive technologies. This is reflected in the book *Engines of Creation*, aimed at a broad general audience, and the founding of the Foresight Institute. At the time of the formation of Zyvex, I was concentrated more on the latter, on societal issues and transformation and advanced computation and social software that might address these issues." Besides, adds Drexler, "Zyvex was in good hands under the direction of Ralph Merkle."

Jim did hire – not as a director or manager, but as a consultant – Drexler's co-conspirator, Ralph Merkle. He also hired another of Drexler's Foresight colleagues, Robert Freitas, who is notable as the author of *Nanomedicine* [14], a multi-part prospectus (some would say fantasy) on the application of molecular nanotechnology to the practice of medicine. In fact, while he worked at Zyvex, Freitas' major activity was the writing of *Nanomedicine*, a very dense, well-researched treatment of the subject – even if some of Freitas' notions are a little over-the-top.

Jim and Zyvex quickly found that creating a nanotech assembler from scratch was a daunting task. Research at the company descended into a sort of academic model, or perhaps a Xerox PARC model, with projects and people proliferating. When Tom Celluci was made President and COO in 2003, he says, there were "... 82 employees and 85 projects." He cut the number of projects to three. The dream of a nanotech assembler was put on hold while the company concentrated on projects that had a more immediate prospect of commercialization. Employees who couldn't respond to Celluci's mantra, "Better, faster, cheaper," were invited to leave.

Among those who left Zyvex was Merkle, eventually finding an academic position in the Computer Science department of Georgia Tech. Celluci complained that Merkle's interviews with the press about the company would always descend into discussions of life-extension and freezing heads, which in Celluci's view undercut the serious image that Zyvex hoped to project. Jim Von Ehr said that Merkle seemed interested in designing all the wonderful things that would be possible once a molecular assembler was available, but wouldn't concentrate on the assembler itself. "It was like imagining all the places you could go if you had a faster-than-light spaceship," says Jim. But first, of course, have to build the spaceship.

Zyvex's mission, as currently stated on its website, "... is to become the leading worldwide supplier of tools, products, and services that enable adaptable, affordable, and molecularly precise manufacturing." But Jim insists that the long-term goal is still the elusive molecular assembler. His assembler is nothing like the replicating nanobots in the *Engines of Creation*. So far, he points out, "We haven't built an autonomous robot that can walk around the block and find its way back home. A fruit fly has more sense than any of the robots built so far." Instead, the Zyvex approach is more like a nanoscale Henry Ford assembly line driven from above by software and computers. Just as robotic arms manipulate parts to assemble cars, so would miniaturized machines handle nanoscale parts to make devices with nanoscale architecture.

The main thrust of Zyvex's business, so far, is the manufacture and sale of nanomanipulators that allow the gripping and positioning of nanoscale objects. These are not yet molecular assemblers, but they have been able to grab and position objects as small as 10 nm in diameter. Zyvex also has a microelectromechanical systems (MEMS) project funded by DARPA that it hopes will be the forerunner of the ultimate assembler. MEMS are microscale machines that are constructed in three dimensions, primarily through the use of lithography (see Chapter 3).

Zyvex has also become adept at manufacturing and handling carbon nanotubes. One of the problems with using nanotubes has been their tendency to aggregate. Zyvex has engineered a molecule with an affinity for carbon nanotubes that disperses them in solution. A second part of the molecule effects the anchoring of the nanotubes in epoxy. This allows the much stronger carbon nanotubes to replace carbon fiber in various applications. Zyvex is supplying its NanoSolve carbon nanotube material to Easton Sports, which makes bicycle parts.

Jim Von Ehr has plans beyond Zyvex and the assembler. Listening to him, I had to keep reminding myself that a successful technology entrepreneur has to be almost pathologically optimistic in order to bring an idea – an insubstantial thing that is little more than the ordered firing of neurons – into concrete reality. Otherwise, he could not survive the ups and downs of business and the markets over which he has little control. By being essentially the sole funding source for Zyvex, Von Ehr has spent more of his own money supporting nanotech than any other single individual on the planet.

Along with Richard Smalley, Von Ehr was one of the few nanotech pioneers that were invited to the White House to watch President George Bush sign the 21st Century Nanotechnology Research and Development Act, which promises $3.7 billion in federal funding for the young industry. Jim kicks himself now that he did not use the opportunity to bring up the president what he thinks is the most important "grand challenge" for nanotechnology: the replacement of fossil fuels as an energy source. How would this work? Well, he is a little vague on that – he has some ideas kicking around – thermal energy conversion to electricity, for instance, but that may be the subject of some future interview. I offered the opinion that it can come none to soon, with the Earth already heating up from the build-up of carbon dioxide. Not to worry, says Jim, nanotech will take care of that.

Jim is very impressed by trees, which use sunlight to catalyze the conversion of carbon dioxide to cellulose. Surely if trees can do it, we can do it too. We could, he imagines, create nanotech-enabled carbon dioxide scrubbers that would allow us to determine the setpoint of carbon dioxide in the atmosphere. The question we need to ask is "What temperature do we want the world to be?" He suggests that the optimum temperature will be a subject of negotiation between those living in Siberia-like climes, those who live at the equator, and everybody in between. It is time for us to seize deliberate control of the atmosphere and the climate. The verb "to terraform" arises out of science fiction, and is the process by which one converts a planet that is hostile to life into a planet that is more like Earth. Jim thinks it is time to terraform Earth itself.

Meanwhile, there is Zyvex, which he says is on course to be cash flow-positive in 2006. Maybe an initial public offering will give Jim the wherewithal to pursue his ever-more ambitious goals.

Ernst Ruska and Gerd Binnig

The nanomanipulators that Zyvex sells are engineered to work in consort with electron microscopes and scanning probe microscopes, which now allow imaging down to the sub-nanoscale. This means that individual atoms can be seen. There was a time, not long ago, when scientists would have considered this an impossibility.

In a sense, microscopy goes back a very long way (Table 4). The ancient Egyptians recognized the phenomenon of magnification by glass, although they didn't seem to understand that magnification was an intrinsic property of the *shape* of the glass. So it wasn't until 2500 years later that the first lenses were made. Two hundred years later still, someone thought of mounting them over their nose to make a pair of spectacles.

Optical microscopes, in very much like their present form, were available by the 17th century. They were good enough already to see protist micro-organisms or "animalcules" as they were described by Anton van Leeuvenhoek, one of the earliest microscopists. Leeuvenhoek was also credited with the discovery of sperm, as is illustrated by his own drawing, apparently of his own sperm (Fig. 14).

Table 4 The history of microscopy.

1500 BC	Magnification recognized
1000 AD	Lenses
1200	Spectacles
1590	Optical microscope
1667	Compound microscope
1679	Reflecting microscope
1933	Electron microscope
1940	Transmission electron microscope
1953	Scanning electron microscope
1981	Scanning acoustic microscope
1987	Atomic force microscope

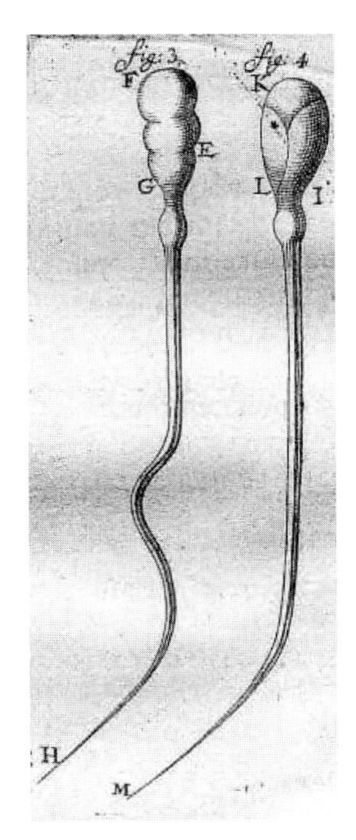

Figure 14 Human sperm, as drawn by Anton von Leeuvenhoek.

Although microscopes and imaging became refined over the next three hundred years, there was no real conceptual advance until the 1930s, when Ernst Ruska and Max Knoll invented the electron microscope. It was around this time that the quantum nature of matter was being elucidated by Niels Bohr, Albert Einstein, and Max Planck among others.

Ruska did not understand that electrons were wavelike when he began his work on the electron microscope – but he did know that the resolution of the optical microscope was limited by the wavelength of light. Ruska figured that electrons, like little bullets, would travel in straight lines. So he was a little disappointed when the quantum physicists showed that electrons also travel in waves.

Fortunately, de Broglie was soon to show that the wavelength of electrons were 100 000 times smaller than the smallest wavelengths of light. This was small enough to allow a 2.2 Angstrom level of resolution, calculated Ruska and his colleague Max Knoll in 1932. This level of resolution was not actually reached until forty years later, and sub-angstrom resolution was actually achieved in the 21st century. In the meantime, the electron microscope did give an unprecedented window into the world of the very small (Fig. 15). Materials scientists and particularly biologists made use of this new tool. By the 1970s, researchers were able to visualize individual proteins and DNA molecules.

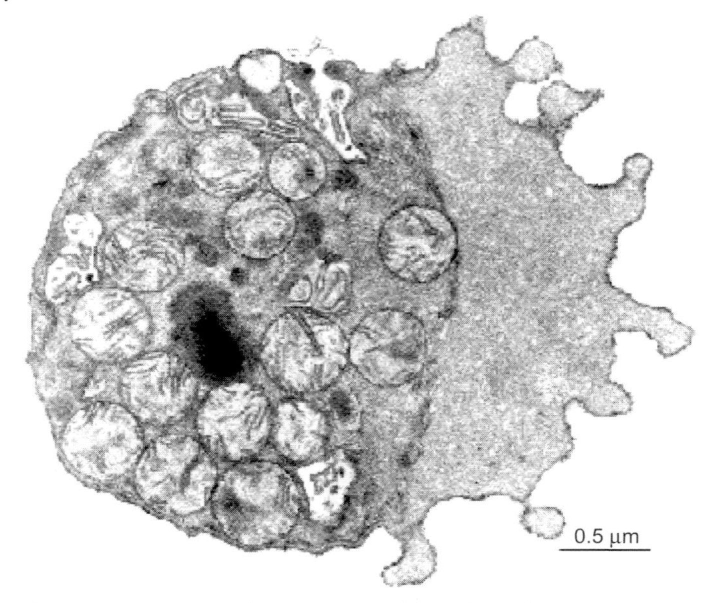

Figure 15 Transmission electron micrograph of a sperm cell from *Caenorhabditis elegans*, a type of worm, seen in cross-section. The round organelles are mainly mitochondria, whereas the dark dense spot in the middle is the nucleus containing condensed DNA. Instead of a tail for swimming, *C. elegans* sperm have pseudopods for walking, as seen the right-hand side of the image. From Sam Ward, of the University of Arizona.

Like an optical microscope, an electron microscope has a focusing lens, except that the focusing is carried out magnetically rather than with a piece of glass. Ruska reported in his Nobel address that he had once discarded the magnetic lens on theoretical grounds, but continued to work with it anyway for lack of a better solution. Sometimes, the slow slog works better than brilliant insight!

As people cannot see electrons, visualization with the electron microscope is indirect. It creates a picture pixel-by-pixel, very much like a television set. In fact, Max Knoll left off work on the electron microscope to take a position with Telefunken of Berlin to work in the developing field of television.

Much of Ruska's early work on the electron microscope was performed as an unpaid postgraduate fellow; fortunately for the world at large he was unable to obtain a reasonable paying job in economically downtrodden Germany between the World Wars. Fortunately for Ruska, he was long lived – he didn't collect the Nobel prize for his invention until 1986 at the age of 80, two years before his death (Nobel prizes cannot be given posthumously). He shared that prize with Gerd Binnig and Heinrich Rohrer, from IBM's Zurich laboratory, who collaborated to create the first scanning tunneling microscope (STM). It was this instrument that allowed the first "visualization" of individual atoms. The instrument actually measures the transfer of electric charge between a very fine probe tip and

Figure 16 Ernst Ruska with Nobel Prize medal. Reproduced courtesy of The Nobel Foundation.

a sample as the probe is scanned over it. Because of this, that microscope (actually a nanoscope or even a femtoscope) only works on samples with conductivity.

There was widespread disbelief among many that the STM could actually image individual atoms; Binnig and Rohrer were even accused of cheating by using computer simulations. Since they worked for Big Blue, there was perhaps an understandable paranoia on the part of academic scientists that the two had access to unlimited computer power. Perhaps because of this, in their first major paper using the new instrument, Binnig and Rohrer showed their results in a very visual, non-computational manner. Tracings from the scans (similar to those of an oscilloscope) were traced out on plastic sheets; these were then cut out and glued together to make a composite that demonstrated the relief from the sample. The hills and valleys of this three-dimensional plastic topo map illustrated positions of individual atoms in their sample of silicon.

Radically new science undergoes a transition from (1) outright derision, to (2) grudging acceptance that the experiment has merit, to (3) acceptance of the result

Figure 17 Gerd Binnig. Courtesy of the Nobel Foundation.

as a classic in the field suitable for inclusion in textbooks, and finally to (4) eventual dismissal as "old hat." The STM made this journey in record time, from invention in 1981 to the subject of a Nobel prize a mere five years later. A year earlier, Binnig had an inspiration for the atomic force microscope (AFM), an even more powerful instrument that has largely displaced the tunneling microscope as a research tool.

The story goes that Binnig was lying on the sofa staring at a stucco ceiling, imagining the bumps on the ceiling as individual atoms, when he suddenly realized that he could devise a means of imaging each bump directly, without the need for electron tunneling. Within only a few weeks, he had turned his idea into the prototype of the AFM with help from Calvin Quate, a diverse genius from CalTech. Their invention is best understood as a variation on the diamond needle and turntable used to play vinyl records in the days before CDs. In this case, the stylus is sharp to within a few atoms. As it is dragged over the topography of the sample, the deflections of the needle are recorded as an amplitude tracing. The sample is scanned by the needle at intervals measured in fractions of a nanometer, giving an exquisitely detailed three-dimensional map of the surface.

Although the concept of the AFM is disturbingly simple, the implementation is not. Obviously, vibration control is a serious problem. Without it, imaging atoms with an AFM is somewhat like playing an old Victrola while bouncing through the Rockies on a jeep trail.

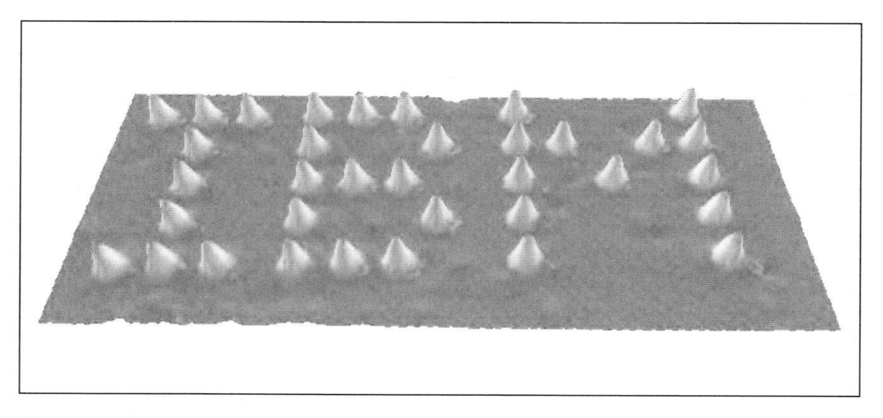

Figure 18 "The Beginning", created by Don Eigler from 35 xenon atoms on a nickel surface with a scanning tunneling microscope. The apparent cone-shape of the xenon atoms is an artifact caused by the shape of the scanning probe tip. Picture used courtesy of IBM Corp.

The AFM and the STM, and the many variations of scanning probe microscopes that were eventually invented, can be used for more than simply observing the nanoworld. They serve as an interface through which some aspects of the nanodomain could be impacted and you could simultaneously gain feedback on the results of your manipulation. It was actually with a hand-made STM that Don Eigler pushed xenon atoms around on a nickel surface to spell out the letters IBM in 1989, allowing Big Blue to be the first to get its logo down to nanoscale. In the words of journalist Alison Overholt from fastcompany.com, the STM was "... cobbled together from odds and ends – cable and wiring encased in half of an Arizona Iced Tea can, vacuum chambers wrapped in tinfoil to help retain heat – the [instrument] seems more like a Calvin and Hobbes Transmogrifier than an ultra-sensitive tool that creates nanoscale structures from individual atoms." Nevertheless, it worked.

This image of the IBM logo that Eigler created hangs in a gallery at IBM's Almaden Research center, where it is appropriately titled "The Beginning" (Fig. 18). And indeed, this image started the nanoage in earnest. Atoms stopped being a kind of theoretical construct that we learned about in physics class, and became real, tangible bumpy objects that we could visualize and manipulate at will, albeit not without considerable effort, skill, and technology. Ten years later, Lee and Ho demonstrated that an AFM tip could be used to push atoms together to form covalent bonds, and demonstrating that mechanical molecular assembly was feasible. On this point, Drexler has been vindicated, although practical application of mechano-synthesis to build useful objects is still a long way off.

Mike Roco

Most nanotech developments during the twentieth century, for instance AFMs, carbon nanotubes or dendrimers, occurred through the research efforts of large corporations. But even IBM, the most nanotech-savvy of the multinationals, didn't officially designate nanotechnology as a priority until 1999. Until the 21st century, government financing for nanoscale research was incidental at best, but this began to change in November 1996 when Mike Roco, of the National Science Foundation (NSF), convened a small group of researchers to plot out a long-term strategy for what eventually became known as the National Nanotechnology Initiative.

Roco came to the NSF in roundabout fashion after previous career stops in Delft, Tokyo, Kentucky, and Caltech. Though he prefers the strong, masculine Anglo appellation "Mike," suitable for American truck drivers and linebackers, his

Figure 19 Mihail "Mike" Roco of the National Science Foundation. Courtesy of Mike Roco and the National Science Foundation.

true first name is Mihail, and he was born in Romania. His father's family was of Italian extraction and his mother came from Paris. He admits to speaking several languages. He argues that his former life as an international man of mystery is irrelevant to his status as an American government official. He seems to fear that Americans will find him too foreign. Not to worry. A nation that welcomed Zbigniew Kazimierz Brzezinski as its National Security Advisor can accept Mihail Roco.

Roco speaks fluent, though heavily accented, English. His words come out in whole paragraphs. One can practically see the bullet points as his mind tracks along well-worn grooves laid down in innumerable policy discussions, debates, and seminars that he has participated in over the past few decades as Prime Nanotech Emissary/Missionary to the world.

Roco is an engineer by training. His interest in nanotechnology was kindled when he worked on an industrial project involving ultra-small particles. These just didn't behave the way he expected them to. Science, he points out, can predict the properties of bulk materials very well. There is also a pretty good theoretical understanding at the atomic level, thanks to chemistry and atomic physics. It is the intermediate level, when dealing with clusters of atoms – the macromolecular scale upon which life itself is founded – that our intuition fails us. Quantum effects, surface effects, and the relatively weak Van der Walls forces cause matter to behave in ways that are weird, wonderful and sometimes downright spooky.

Roco's fascination with nanoparticles led him to write a grant proposal to NSF for a project under an emerging technologies initiative. Roco came to the NSF to head the $3 million project in 1990. Roco's relatively modest effort came to be the seed crystal for the much larger National Nanotechnology Initiative.

In 1996, Roco assembled an informal group to discuss strategies for creating a national strategy for nanotech research. Among the people involved were Paul Alisatvos (University of California Berkeley), Stan Williams (Hewlett Packard) and Jim Murray (U.S. Naval Research Laboratory). The group went about it in very systematic fashion. Roco points out that, at the nanoscale, manufacturing processes are not unique to any one industry. When you are using bulk materials, you are working like a sculptor, trying to liberate, for instance, the inner car body locked within a mass of sheet metal. Your experience in building cars is not much help if you want to fashion mahogany cabinets. When you are working from the bottom-up, molecule-by-molecule, the same processes that allow you to make a memory chip may allow you to build a biosensor or a designer drug.

The initial group of hopeful researchers came up with ten areas of relevance where nanotechnology could benefit academic science, government and industry. From this framework, they began soliciting intellectual contributions from other experts; eventually over 150 researchers contributed to the final National Nanotech Initiative (NNI). Roco's group was remarkable in the sense that a lot of effort was expended before there was any real expectation of funding. The NNI was initially a kind of prospectus; a red herring optimistically prepared in the belief that it could by its powers of persuasion obtain a constituency among government funding agencies.

A window of opportunity opened in 1999. The U.S. economy during the 1990s had exceeded all expectations, driven in part by the remarkable emergence of the Internet. As a consequence, the Clinton Administration was running a budget surplus (soon to go down the drain in the Bush Administration, but that is another story). A surplus in the American government is unheard of in modern times; it was burning a hole through the pockets of politicians and bureaucrats. Clinton wanted to hear about useful ways to spend it. Fortunately, Roco had already made contact with White House economic advisors, and so on March 11, 1999, Roco was summoned to the Indian Treaty Room in the White House to give a presentation. A more unlikely setting for an examination of high technology is scarce to be imagined. Originally, it was built as the Navy Department Library and Reception Room. A rococo masterpiece of Italian and French marble and cast-iron railings, the room features nautical motifs, such as shells, seahorses and dolphins. The Navy had moved out by 1921. Nobody now remembers how or why this space was renamed the Indian Treaty Room long after most of those treaties had been signed, broken and forgotten.

Andy Warhol said that in the future, we would all be given 15 minutes of fame. Roco was given only 10 minutes to make his mark in front of the assembled economic and science advisors to the President of the United States. He used it to advantage, making a case for nanotechnology and requesting a mere $500 million in start-up money. Many questioned the price tag, but eventually, Roco notes with pride, the first year of funding under the NNI came out to $490 million. But it wasn't an easy accomplishment. There were serious political minefields to negotiate.

Bill Clinton formally announced the NNI in January of 2000, when he was at the end of his two terms. The 2000 election was a particularly bitter fight between Clinton's vice president Al Gore and Texas governor George W. Bush. Gore received the majority of the popular vote, but the virtual tie in Florida left the electoral vote undecided until the Supreme Court made a split-decision in favor of the Republicans. George Bush came to office with no popular mandate, but nonetheless very different priorities from his predecessor. The picture was complicated in short order by an economic downturn and the most audacious and successful terrorist act ever perpetrated, when the twin towers of the World Trade Center collapsed in New York on September 11, 2001, killing almost 3000 people.

Mike Roco emerged from the Clinton administration with a new title, Chair of the Subcommittee on Nanoscale Science and Engineering for the National Science and Technology Council, of which the President of the United States is titular head. Roco's subcommittee is the only one that remained unchanged between the two administrations. Using a nanoscale lever and his Chair as a place to stand, Roco was able to move the world down the road to its nanotech future.

Roco seems to have taken to heart the statement attributed to Alan Kaye of the fabled research lab Xerox Parc, "The easiest way to predict the future is to create it." No one has been more active in pushing the nanotech future than Mike Roco.

In 2001, Roco and his NSF co-conspirator, William Bainbridge, released a controversial assessment claiming that nanotechnology would amount to a one-trillion dollar industry by 2015. Four years later, nanotech accounts for about $60–70 million in revenues by Roco's own generous estimate. Still, Roco stands by the initial trillion dollar claim. In his defense, it should be noted that some independent market research firms have come out with even higher estimates. Lux Research, for instance, has publicly estimated that products that incorporate nanotech will account for $2.6 trillion in sales by 2014, accounting for 15% of the global economy. Numbers, however, can be misleading: because your Cadillac incorporates a nanotech sensor in its fuel injection system, does this mean that the whole Cadillac is a nanotech product?

Roco was instrumental in setting up a collaborative project between the NSF and Department of Defense aimed at understanding the convergence of technologies at the nanoscale. A workshop was held in December 2001. It was the consensus view of the participants that nanotechnology, biotechnology, information technology and cognitive science were on the glide path to convergence into one grand unified scheme, with potentially enormous consequences for future industry and society. In June 2002, the group published a report titled *Converging Technologies for Improving Human Performance* [18] relating how "Nano-Bio-Info-and Cogno" were becoming one integrated discipline. The report called for a "New Renaissance" of science and society based on technologies united by "material unity at the nanoscale."

Some of the NBIC schemes had a decidedly Borg-like quality – for instance, the creation of a brain/machine interface that would allow operation of machinery (jet fighters, say) by simply thinking, speeding up reaction times. The highest priority was given to the "Human Cognome" project, designed to understand and if possible, enhance the function of the human mind, including interfacing with new sensing modalities (five is not enough). Given the Defense Department sponsorship, there was naturally a heavy dose of military applications for NBIC technologies, including ubiquitous sensors and robotic weapons systems. Some thought was given on the need for society to re-organize in response to pressures brought to bear by NBIC convergence. For instance, this passage from the Overview of the report:

> People will possess entirely new capabilities for relations with each other, with machines, and with the institutions of civilization. In some areas of human life, old customs and ethics will persist, but it is difficult to predict which realms of action and experience these will be. Perhaps wholly new ethical principles will govern in areas of radical technological advance, such as the routine acceptance of brain implants, political rights for robots, and the ambiguity of death in an era when people upload aspects of their personalities to the Solar System Wide Web [18].

If the NBIC report had been put out by the Drexler's Foresight Institute, it would probably have not been given much weight. But coming from two powerful agencies of the United States government, it had an enormous impact, not all of it positive. Many people are not comforted by U.S. military involvement in nanotech development, for instance. For others, the NBIC reinforced the science fiction aura surrounding nanotech. The issuance of the NBIC report, however, put the

U.S. government firmly in the believers' camp with respect to the transformational powers of technology at the nanoscale.

By 2002, when the NBIC report came out, Japan, Korea and China, had begun working on their own nanotech initiatives. This was probably not a coincidence. Roco, himself, spoke before committees working on national programs in Japan, Korea, Australia and the European Union. By 2005, over 62 countries had some sort of national nanotech effort. The funding of nanotech research by governments around the world springs forth not only out of the spirit of international competition but also of the fear of being left behind technologically, economically and militarily.

In the U.S., the National Nanotech Initiative led to Twenty-First Century Nanotechnology R & D Act, which was passed in the House by a bipartisan vote (one of the few) 405–19 and in the Senate unanimously. The act was signed into law by President Bush on December 3, 2003, and carried a $3.7 billion price tag. Almost immediately, the Japanese government reportedly pledged to match the U.S. dollar for dollar, or yen equivalent thereof.

Though Mike Roco has had a fine career as a mechanical engineer, his main legacy with respect to nanotechnology will be through the influence he brought to bear in persuading the U.S. government, not to mention the rest of the world, to provide funding. It has been a masterful job of engineering the future.

References

1 Feynman, R. and Leighton, R., Surely you're joking Dr. Feynman: Adventures of a Curious Character, W.W. Norton and Company, **1985**.

2 Feynman, R. There's Plenty of Room at the Bottom, Engineering and Science 20: 22 (February), **1960**. Reprinted in Miniaturization, edited by H. Gilbert, Reinhold, **1961**.

3 Watson, J. and Crick, F., A Structure for Deoxyribonucleic Acid, Nature: 171: 1737 (**1953**).

4 Drexler, K.E., Molecular engineering: An approach to the development of general capabilities for molecular manipulation. *Proc. Natl. Acad. Sci. USA* 78: 5275–5258 (**1981**).

5 Drexler, K.E. Engines of Creation: The Coming Era of Nanotechnology, Anchor Books, **1986**.

6 Drexler, K.E., *Nanosystems: Molecular Machinery, Manufacturing, and Computation*. John Wiley & Sons, Inc.: New York, Chichester, Brisbane, Toronto, and Singapore (**1992**).

7 Edwards, S.A., B-162 Biomedical Applications of Nanoscale Devices, Business Communications Co., **2003**.

8 Crichton, M. Prey, HarperCollins, New York, **2002**.

9 Kurzweil, R. and Grossman, T., Fantastic Voyage: Live Long Enough to Live Forever, Rodale Press, **2004**.

10 Kurzweil, R., Law of Accelerating Returns, www.kurzweilai.net/articles/art0134.html.

11 Edwards, S.A., Mind Children, 21C: Scanning the Future, No. 23.

12 Joy, W., Why the Future Doesn't Need Us, Wired 8.04, **2003**.

13 Smalley, R., Of Chemistry, Love, and Nanobots, Scientific American, September, **2001**.

14 Baum, R., Drexler and Smalley make the case for and against 'molecular assemblers'. Nanotechnology 81: 37–42 (**2003**).

15 Freitas, R. (**1999**), Nanomedicine Vol. I: Basic Capabilities, Landes Bioscience;

Nanomedicine Vol. IIA (**2003**), Biocompatibility, Landes Bioscience.

16 Atkinson, W.I., Nanocosm, Amacom: American Management Association, **2003**.

17 Lee, H.J. and Ho, W., Single bond formation and characterization with a scanning tunneling microscope. Science 286: 1719–1722 (**1999**).

18 Converging Technologies for Improving Human Performance, edited by M. Roco and W. Bainbridge. A report prepared by the National Science Foundation and the Department of Defense. www.nano.gov.

Chapter 3
On the Road to Nano-

The nanoscale is a destination approached by a number of different paths. Although visionaries such as Feynman, Drexler and Roco imagined in detail how nanotechnology might be accomplished, actual workable nanotechnology has arrived organically, as an outgrowth of work in biology, chemistry and semiconductor manufacture. In principle, much of the technological progress of the twentieth century might be seen as a necessary precursor to nanoscale engineering. Nanotech instruments such as the atomic force microscope (AFM) could not have been built without the prior development of the precision fabrication techniques used to manufacture the device. Without precision engineering required to control the movement of an AFM tip within subatomic tolerances, for instance, the instrument would be of little use. Computer science and software are also necessary to translate its tracings into a two- or three-dimensional images. In this chapter we will concentrate, however, on three developments that were critical to the emergence of nanoscience: lithography; molecular biology; and supramolecular chemistry.

Lithography

If we think about the term "lithography" (from the Greek for "stone writing"), most of us probably think in terms of the art world; an Escher print is an example of a lithograph. The process was actually invented by a playwright named Alois Senefelder in 1798. In its original implementation, an image was painted onto limestone with grease, and the stone was then dipped in ink. The grease would retain the ink while the rest of the stone would repel it. The stone was then pressed onto a piece of paper to create a print.

Lithography is also used to make semiconductor chips. In this case, the "stone" is the silicon from which chips are made. The silicon is written upon by irradiation with a light source through a mask (like a stencil) to make a particular pattern. The pattern is then etched into the surface using a chemical developer. The radiation changes the reactivity of the silicon to the developer.

The Nanotech Pioneers. Steven A. Edwards

ISBN: 3-527-31290-0

The smallest width of the pattern that is obtainable through lithography is limited by the wavelength of light. So far, the visible spectrum has been used for semiconductor manufacturing. However, to make chips with features below the current standard of 157 nm, it will be necessary to use smaller wavelengths. A number of strategies have been suggested, but it appears at present, that the industry is coalescing around "extreme UV", which is actually in the X-ray spectrum. Electron beam lithography, which uses a stream of electrons instead of light, can produce a pattern with a resolution as small as 10 nm, but is not well-suited to mass production.

Although lithography has been used primarily to make two-dimensional circuits, in recent years it has become possible to sculpt three-dimensional objects out of silicon using specialized forms of lithography. Tiny gears and even working motors have been made this way. These devices are known collectively as micro-electromechanical systems. This multisyllabic term is usually abbreviated as MEMS (rhymes with stems) to save on verbiage.

MEMS have made it to the commercial world in the form of various sensors. An accelerometer, for instance, is a MEMS device that senses rapid changes in motion and generates an electronic response. Accelerometers have found use in air bags that protect us in automobile collision. They are also found in heart pacemakers – they tell the heart to speed up when more energy is needed during physical exercise.

Although MEMS are very small, they are still within the micro range, one hundred to a thousand times large than nanoscale objects. To make the nanoscale machines dreamt of by the nanotech promoters, new means will have to be found.

On start in this direction is the "soft lithography" or "nanoimprint lithography" invented largely in the Harvard laboratory of George Whitesides. A mold of hard material is etched by electron ion beam lithography. This is then pressed into a soft rubber-like polymer to make an imprint. The raised portions of the imprint can then be coated with molecular ink and used like a rubber stamp. The original version of this technology was called microimprint technology, but refinements have carried it down into the nanoscale.

A variation on Whitesides' technique is being commercialized by a small entrepreneurial company called Molecular Imprints in Austin, Texas. They call their technology Step and Flash Imprint Lithography (S-FIL). S-FIL uses a fused silica template with a circuit pattern etched into it. This surface is covered with a release layer, and then pressed into a thin layer made from a type of low-viscosity polymer containing silicon. The surface polymerizes to a hard layer when irradiated with an ultraviolet lamp. When the template is removed, the circuit pattern is left on the surface. A residual layer of polymer between features is eliminated by an etch process, and a perfect replica of the pattern is ready to be used in semiconductor processing The limit of resolution is set by the technique used to create the initial template, which is done currently with an electron beam writer. Molecular Imprints claims to have pushed resolution down to 20 nm. The company has already begun shipping its lithography products to Motorola and other semiconductor manufacturing companies.

Another small company, NanoInk, founded by Northwestern University's Chad Mirkin, is commercializing a technology called "dip pen nanolithography." This technique looks more like writing with Thomas Jefferson's quill pen than it does conventional lithography, except that the quill pen has been shrunk to where its point is a few nanometers at its tip. In actuality, the pen is essentially the probe tip of an AFM. The precision movements of the scanning scope are used to create lines of precise dimension, down to 10 nm in diameter, so far.

Dip pen lithography has potential applications both in electronics (imagine drawing very small circuits) or in creating very small arrays used in biochemical assays. In the latter case, the "ink" might be particular DNA oligonucleotides corresponding to individual messenger RNAs in a cell. Thousands of assays could be carried out in a space the size of the period following this sentence. Such extreme miniaturization has advantages in the conservation of resources and offers the advantage of performing detailed assays on the contents of a single cell.

Erez Braun and colleagues at Technion-Israel Institute of Technology have created a process that they call "sequence-specific" molecular lithography. This process actually uses DNA to create networks and junctions. The DNA is coated with particular metals in a sequence-specific manner, and the metal is then used to localize molecular objects at a specific "address" based on the DNA sequence. The system has been used, for instance, to create a field effect transistors, each composed of a protein and a carbon nanotube.

In his lecture, "There's Plenty of Room at the Bottom," Feynman [1] imagined a top-down process in which miniature tools were used to make even smaller tools, which were used to make yet smaller tools, until we reached the stage where tools could be employed to move atoms around one at a time. So far, lithography is the best embodiment of this kind of top-down process, in which the limit of resolution keeps getting pushed downward. We are now able to fabricate tips for AFMs that come to a nearly perfect point of one or at most a few atoms, and experimental electronic components made of molecules.

Richard Feynman pointed out that "... if we go down far enough, all of our devices can be mass produced so that they are absolutely perfect copies of one another. We cannot build two large machines so that the dimensions are exactly the same. But if your machine is only 100 atoms high, you only have to get it correct to one-half of one percent to make sure the other machine is exactly the same size – namely, 100 atoms high!"

We're not there yet, but we're closing in.

Molecular Biology

Biotechnology, a commercial adaptation of the science of molecular biology, is sometimes referred to as "wet nanotechnology" or, by some wags, "nanotechnology that works." As technologies go, this is a very young one – it did not begin in earnest until the last half of the twentieth century. Fortunately, it builds on devices perfected by Nature over 3.5 billion years of evolution. "Evolution has produced an overwhelming number and variety of biologic devices, compounds and processes that function at the nanometer or molecular level and that provide performance that is unsurpassed by man-made technologies," notes Raj Bawa [2], founder and president of Bawa Biotechnologies Consulting, LLC.

The nascent field of molecular biology was impacted by a number of European physicists, many of them dismayed by the development of the atom bomb, who were looking for new field that was not, in their view, morally compromised.

The great physicist Erwin Schrödinger, famous for his uncertainty principle, was fascinated by life as it seemed to contradict the laws of physics and thermodynamics, as they were understood in 1944. The three laws of thermodynamics were particularly troublesome: (1st) energy cannot be created or destroyed; (2nd) entropy (or disorder) always increases; and (3rd) entropy is zero only in a perfect crystal at a temperature of absolute zero (which cannot be obtained). These are sometimes summarized as follows: you can't win; you always lose, and you can't even quit the game (or sometimes, Mother Nature is a bitch).

In living and replicating themselves, organisms use energy to create ordered copies of themselves that increase in number exponentially, given enough food and freedom from predators. Though biochemists and molecular biologists can now show how life actually exploits thermodynamic laws to achieve its miracles, this was far from obvious to a physicist in the early part of the twentieth century.

"How can we from the point of view of statistical physics," asked Schrödinger, "reconcile the facts that the gene structure seems to involve only a comparatively small number of atoms ... and that nevertheless it displays a most regular and lawful activity with a durability or permanence that borders upon the miraculous?" [3]. A student of Schrödinger's, Max Delbruck, who was to become one of the founding fathers of molecular biology and a Nobel laureate, speculated that the gene was an aperiodic crystal or a solid, else how to explain its permanence from generation to generation? Delbruck's intuition was sound – although DNA is not a crystal as it exists in the cell it is aperiodic and, like a crystal, a very stable molecule.

By 1952, it had become accepted on the basis of bacterial experiments that the gene was carried within molecules of DNA. In 1953, Watson and Crick put forth their model of DNA structure, the double helix. As each strand of the double helix was a mirror image of the other, essentially a copy of the same information, it became readily apparent how the genome could be duplicated without being destroyed. "It has not escaped our notice that the specific pairing we have postulated immediately suggests a possible copying mechanism for the genetic material," remarked Watson and Crick, laconically in their landmark paper. The details of that replication, however, proved surprisingly complicated, and required more than a decade to work out in any detail.

The same decade was marked by frenzied activity in which the role of mRNA was elucidated, the relationship between DNA sequence and protein sequence determined, and the broad outlines of ribosome function in the synthesis of proteins established.

By 1963, according to Gunther Stent [4], molecular biology had entered the "academic phase" with all of the fundamental problems already solved – a dead field.

"All hope that paradoxes would still turn up in the study of heredity had been abandoned long ago, and what remained now was the need to iron out the details." This is a classic example of the dumb statements that otherwise brilliant scientists are inexplicably inclined to make.

Jim Watson's classic textbook, *The Molecular Biology of the Gene* (which inspired me, among others to become a biologist), reported on "the Central Dogma" of molecular biology – information flowed from DNA to RNA to protein. Another item of the dogma was that each gene coincided with one protein and that there was a colinear relationship between information in DNA copy of the gene, the sequence of RNA bases in messenger RNA, and the order of amino acids in encoded protein.

The Central Dogma developed a serious crack in 1970 when David Baltimore and Howard Temin, working independently, discovered a viral enzyme called reverse transcriptase. Reverse transcriptase is responsible for transforming the RNA genome of a group of RNA viruses now called retroviruses, into a DNA copy that could be integrated into the genome.

The family of retroviruses includes HIV, the virus that causes AIDS, as well as several human viruses that cause leukemia. Unlike the DNA polymerases that replicate cellular DNA, reverse transcriptase is hopelessly error-prone, which is one reason that it is so difficult to create an HIV vaccine. Virtually every copy of HIV genomic DNA contains at least one error. As a result, a huge repertoire of slightly variant HIV viruses accumulates in the blood and body fluids of an HIV-infected patient. Since every viral protein has many variants, it is very difficult to develop a vaccine that will work on all of them. Also, the use of anti-HIV drugs effectively selects for whatever HIV variants are resistant to the effects of the drug. As time goes on, HIV becomes more and more like a collection of virus species rather than just one.

Eventually, it became apparent that reverse transcriptases were encoded not only in viruses but also within the eukaryotic genome and that, over evolutionary time, many cellular mRNAs had been copied and reinserted at random locations within the genome. These are called pseudogenes, because the vast majority are inactive. Moreover, reverse transcriptase is the ultimate "selfish gene"; endlessly replicating its own message.

Baltimore and Temin were awarded the Nobel prize in 1975 for their discovery of reverse transcriptase. Not long thereafter, I was at a conference in Cold Spring Harbor at an RNA Tumor Virus meeting where Alexander Rich made an unscheduled presentation. This was late-breaking data, still "dripping wet" as molecular biologists use to call it. (Much of data that molecular biologists was in the form of X-ray film, processed by hand in those days. Unwilling to wait for the film to dry to see their results, scientists were constantly holding the newly developed film up to the light, while distilled water dripped down their arms.) Rich's experiments did not even concern RNA tumor viruses, but DNA viruses, instead. It is a measure of the extraordinary importance accorded to the data that Rich was allowed to speak.

Rich and his colleagues had hybridized mRNA encoded by DNA viruses back to the DNA from which they were transcribed (hybridization means that the single-stranded RNA was allowed to form a complex with single-stranded DNA, forming a double helix). The hybridized complexes were then examined under the electron microscope. It was possible to distinguish double-stranded sections from single-stranded in the photograph. What they found was, at that time, astonishing. Large, looped DNA structures remained unhybridized, while the portions on either side of the loops were contained with double-stranded hybrids. This implied that either: (1) the RNA polymerase (an enzyme responsible for making RNA) had jumped from one section of the DNA strand to another in the process of creating the mRNA; or (2) that the mRNA was spliced together from a larger copy that included the looped sections. Eventually, the latter was proven to be true.

Decades of experimentation with bacterial genes had provided evidence for the colinearity of mRNA and DNA. The splicing of mRNA was accepted almost immediately by molecular biologists, because it explained so many problems that they had encountered in dealing with mRNA from eukaryotic cells and their viruses. Later, when bacteria were re-examined, some spliced mRNAs were discovered there too. Eventually, other means of editing mRNAs were discovered. Some RNA bases are actually modified after transcription and so do not accurately reflect the DNA strand from which they were made.

The one gene–one protein hypothesis also had to yield. Experiments conducted by Ron Evans, Michael Rosenfeld and many others showed that the same RNA gene transcript could be spliced in different ways to encode multiple proteins. Thus, while the Human Genome project has indicated that humans have 25 000 to 30 000 genes, the total number of proteins is still anybody's guess.

Francis Collins led the public version of the Human Genome project. A private effort was also undertaken by the biotech company Celera, led by its founder, Craig Venter. Both Collins and Venter are men of colossal ego, and the struggle between them was filled with all the dissension and mud-slinging worthy of a U.S. presidential race. Venter already had a history of antagonizing his academic colleagues stemming from his attempts to patent large swaths of genomic information (the courts have now ruled that it is legal to patent genomic sequences, but only in the context of a particular application). However, Venter is credited with inventing the "shot-gun" method of DNA sequencing that both teams employed. He must also be regarded as something of an organizational genius. The rapid progress made by the Celera team using very fast DNA sequencers prompted a frantic redoubling of effort by the much larger public effort. Eventually, this competition resulted in the completion of the task years earlier than expected, with a "tie" at the finish brokered by none other than President Bill Clinton. Draft versions of the human genome were published by both teams in December 2002. While both sides disparaged the other's efforts in remarkably nasty terms for supposedly dispassionate scientists, the fact remains that their ultimate products were remarkably similar.

The acceleration in the pace of biotechnology has been incredible even to its participants. It took less than 50 years to go from knowledge of the structure of the double helix to the sequencing of the three billion bases of the human genome. It was something that I, for one, certainly did not expect to happen in my lifetime when I received my Ph.D. in biology in 1980. At that time, the sequencing of DNA was arduously slow. The sequencing of a single gene might take years of effort by a hardworking but underpaid postdoctoral fellow. Nanotech-enabled machines may soon make it possible to sequence the entire genome of an individual overnight, turning the process into a standard medical diagnostic test.

The human genome turned out to be surprising in many ways, not the least of which was the number of genes. Prior to the start of sequencing, estimates had been that there would be as many as 150000 genes with a consensus of about 100000, rather than 25000 to 30000 actually found. This is more or less the same as a rat or any other mammal. We are not substantially more complicated, genetically, than a frog or a pufferfish, and we are within an order of magnitude of unicellular organisms like yeast. There are garden plants that have genomes several times the size of ours. One prominent genetic scientist, the founder of a major biotechnology company, publicly refused to believe in the lowered number, arguing that many genes had simply been missed.

The Human Genome Project is now encompassing a variety of "model organisms" including studying the functional genomics many of biologists' favorite pets: *E. coli*, the fruit-fly, yeast, and the nematode. Another project has sequenced the genetic complement of our nearest relative *Pan troglodytes*, the chimpanzee. This is sure to provide another body blow to our *Homo sapiens* ego. Estimates are that the two genomes are 98% identical. Somewhere in the remaining 2% we hope to find the key to all of the things we believe make us unique – language, intelligence, musical ability, or whatever. Comparative genomics may help us figure it out, but there is a burgeoning suspicion that we may not be so unique after all.

Venter has moved on to even greater things; now, he wants essentially to sequence the whole world. In an around the world voyage in his yacht, reminiscent of Darwin's voyage in the *Beagle*, Venter is taking random DNA samples from soil or water. These he sends back to his private institute, The Institute for Genetic Research (TIGR), which sequences them, after which computer algorithms are used to assemble them into individual genomes. Already, many new classes of genes have been found, the functions for which are obscure.

Another Venter project, undertaken in collaboration with Nobel laureate Hamilton Smith, is to develop a "minimal genome" to use for what amounts to synthetic biology. Starting with a very small bacterial genome (an organism called mycoplasma), Venter intends to strip out all of the genes not strictly necessary for replication. Venter will use this to create artificial organisms that he hopes will be useful in creating alternatives to fossil fuels.

The tools and knowledge developing out of molecular biology have been spun into a vibrant biotechnology industry. The most public face of his industry are the biopharmaceutical companies, like Amgen and Genentech, who have supplied the world with new drugs based on human proteins or oligonucleotides (small stretches of RNA or DNA). Beyond this, though, there is another side to biotechnology revolution – industrial biotechnology. Enzymes produced using recombinant DNA technology are used industrially for incorporation into detergents or for making cheese. Chemical companies use genetically engineered bacteria to synthesize organic molecules. Other bacteria are used for bioremediation, to clean up oil spills are other industrial contaminants.

For the nanotechnologists, molecular biology offers a cornucopia of molecular tools and molecular machines. Biological enzymes, for instance, are essentially tiny machines that operate within nanoscale spaces: joining and modifying down organic molecules. Like the mythical "molecular assembler," these devices must be able to "grab" a molecule of a particular shape and chemical composition, maneuver it into position and cause it to react, as necessary, with a second organic compound. In a living organism, this must be done in a very coordinated, regulated fashion depending on physiological needs. Restriction enzymes allow the cutting and splicing of DNA into whatever sequence required. The ribosome, responsible for synthesizing proteins, is essentially the prototype molecular assembler. Programmed by mRNA, it assembles and releases a particular protein type on demand.

It is easy to imagine the biotechnology of today morphing into the nanotechnology of the future. Already there is an effort to develop what is sometimes referred to as the "Mark II" ribosome, capable of synthesizing proteins out of non-natural amino acids. By careful genetic engineering, it is also possible to expand the genetic code to create organisms that will use more than the usual twenty amino acids found in nature. By eliminating redundancy, the three-base genetic code is theoretically capable of encoding as many as 63 different amino acids, with one codon left over to serve as a stop signal.

The world of the living abounds with preformed nanoscale objects that can be easily adapted to the needs of the researcher as he or she attempts to create the new nanotech-enabled Nature of the future. Biomolecules such as protein and DNA are already being exploited to make nanodevices. Like fullerenes, biomolecules generally have a carbon backbone, but one that is widely substituted with modifiable chemical groups. Because biochemistry is well understood, DNA and proteins are being used structurally now to build nanodevices.

Many kinds of biological motors exist, such as the ATP-driven motor that drives bacterial flagella, enabling the organism to propel itself through fluids; kinesin, the monorail which transports intracellular components along microtubule tracks; or the motors that spool DNA and pack it into virus particles. Nanobiologists have even created artificial motors out of biological material, like DNA

DNA is a particularly interesting molecule. Despite the fact that it is only 2 nm in diameter, a single DNA molecule can be up to a meter in length. Truly, this a challenge for nanotechnologists to emulate – how to make something that long that can nonetheless be spooled and made to fit into a cell nucleus. Ma Nature is a pretty incredible engineer, even if she is a bitch!

Supramolecular Chemistry

A molecule is any string of elements connected by covalent chemical bonds – that is, the atoms are bound together by effectively sharing electrons. "Rebuilding the world one atom at time" is the mantra of nanotechnologists who take the bottom-up approach. But one atom at a time is hardly the exclusive province of nanotechnology. Chemists have always done it this way, as they knitted together elements into ever more complicated strings. Until recently, chemistry was done usually in test-tubes or beakers ("Buckets," sneer the molecular nanotechnologists). Molecular reactions in solution, unfortunately are difficult to control – diffusion of the reactants, temperature, pH, aggregation, solvent and surface effects all come into play. "Chemistry," Albert Einstein is supposed to have remarked, "... is too difficult for chemists." Nevertheless, one has to grant that the chemists have made progress. It will take the tiny nanotechnology industry a while to surpass the $2 trillion worldwide chemical industry. Petroleum products, polymers, paint, pharmaceuticals, fabrics, fertilizers, insecticides – these, and many more, are the products that the chemists have given the world.

The world of the biochemist differs from that of the chemist, in that biochemicals exist in continuous interaction with each other, whereas the chemist is usually content to take his or her reactions to a terminal reaction and isolate a product. Or at least that was how it worked before there were supramolecular chemists. Supramolecular chemistry involves the interaction of molecules that are not necessarily covalently bound together – a chemistry that is beyond a single molecule. For instance, supramolecular chemists talk about molecular pairs – a receptor and its substrate – the smaller molecule being the substrate, which is bound to its receptor through non-covalent attraction. The first receptor–substrate pairs were not organic chemicals, however, but proteins. Not only has biochemistry provided supramolecular chemistry with a model to strive for, but the techniques of biochemistry have proved invaluable in creating the large molecular complexes involved in supramolecular chemistry.

Proteins are large, complicated molecules. They have a primary structure determined by their sequence of amino acids, a secondary structure involving localized folding of primary sequence into domains (which may have general features such as an alpha helix or 'pleated sheet'), a tertiary structure determined by the interaction of localized domains with each other, and a quaternary structured determined by the interaction of two or more proteins with each other. Although small peptides are sometimes very stable, most large proteins are not, particularly those with enzymatic activity. The difficulties in working with proteins were recognized early in the science of biochemistry, as suggested this quote from the great biochemist Emil Fischer, circa 1906 [4] (lifted from Bruce Merrifield's Nobel address):

> "Whereas cautious professional colleagues fear that a rational study of this class of compounds [proteins], because of their complicated structure and their highly inconvenient physical characteristics, would today still uncover insurmountable difficulties, other optimistically endowed observers, among which I will count myself, are inclined to the view that an attempt should at least be made to besiege this virgin fortress with all the expedients of the present; because only through this hazardous affair can the limitations of the ability of our methods be ascertained."

Fischer was the first person to chemically synthesize a peptide by joining amino acids together. This was a very complicated procedure that involved forming a peptide bond between the carboxy and amino ends of two amino acids, while blocking any other remaining reactive sites. Then, the product of the reaction had to be purified and crystallized before the chain could be extended. Even modestly sized protein chains, produced in milliseconds by the ribosome, were quite impossible for the chemist until late in the twentieth century. The breakthrough idea, published by Bruce Merrifield in 1971, is so remarkably simple that it seems amazing, in retrospect, that no one had thought of it earlier. Merrifield attached the terminal amino acid to a solid support, actually a polystyrene bead. This made it possible to purify the growing peptide simply by washing away all the soluble chemicals.

Merrifield used his new method to chemically synthesize for the first time a small enzyme called ribonuclease A, which catalyzed the destruction of RNA. Though the yield was not high in these first experiments and the enzyme was never fully purified, Merrifield was able to show that his enzyme had the proper activity. For his efforts, Merrifield was awarded the Nobel prize in 1984.

Merrifield's method of synthesis was easily automated, so that within a few years, there were commercial peptide-synthesizing machines available. All that the chemist had to do was punch in the desired sequence and make sure all of the reservoirs of modified amino acids were kept full. This was a godsend for pharmaceutical companies. Many small peptides have hormonal activities – gonadotropin-releasing hormone (GnRH), which regulates fertility, for instance, is a peptide containing just 11 amino acids. When protein-synthesizing machines became available, it became possible not only to build the natural hormone, but to create analogues in which different amino acids (even non-natural amino acids) were substituted at various positions. This was done with the aim of creating pharmaceuticals that were either more active and more stable than the natural hormone, or which possibly would serve as inhibitors of the hormonal receptor.

Thousands of different versions of GnRH were created using Merrifield's method in hopes of finding the male birth control pill. Alas, this effort was in vain. Although some variants successfully prevented the formation of sperm, unfortunately these also caused suppressed the male sex drive and caused the testes to atrophy. Today, GnRH analogues are used to treat hormone-dependent prostate cancer, premature puberty and other conditions. Many, many other peptide hormones have also been synthesized, including glucagon and somatostatin, both of which are pancreatic hormones involved in glucose regulation, and epidermal growth factor, an important regulator of cell division.

Merrifield's idea led to a whole new field, called "solid-state" or combinatorial chemistry, in which combinatorial polymers, like proteins, nucleic acids, carbohydrates and other organic molecules could be built one unit at a time. Not only was it possible to a build a given molecule with known properties, the technique could also be used to create all combinations of a set of building blocks. Automated DNA synthesis was perfected by a biotech company, Egea Biotech of San Diego, which demonstrated error-free synthesis of DNA strands consisting of thousands of base pairs (Egea was acquired by Johnson & Johnson in 2004). The ability to abiotically synthesize biological molecules is central to a new field called "synthetic biology."

Starting in 1970, using biology as his inspiration, and crown ethers (crown-shaped ring structures of carbon and oxygen) as his starting material, supramolecular chemist Donald J. Cram and his colleagues started to build organic molecules that had biological-type properties. In particular, they looked at "host–guest" complexes – large molecules that might interact with each other in the way that an enzyme might act with a substrate, or that a hormone might interact with a receptor. The early chemists, such as Emil Fischer, had mostly to use their imaginations to visualize what their pet compounds might look like, but by the 1970s Cram had the advantage of fairly sophisticated molecular modeling programs. This is evident in his very visual descriptions [5]:

> Complexes were visualized as having three types of common shapes: (1) perching complexes, resembling a bird perching on a limb, an egg protruding from an egg cup, or a scoop of ice cream sitting on a cone; (2) nesting complexes, similar to an egg resting in a nest, a baby lying in its cradle, or a sword sheathed in its scabbard; (3) capsular complexes, not unlike a nut in its shell, a bean in its pod, or a larva in its cocoon.

Cram concentrated on pairs of compounds that would have the biological-like properties of complementarity (in the way that one strand of DNA is complementary to the other) or "structural recognition" (as an antibody recognizes an antigen). Eventually, he was able to build carbon-based compounds with enzyme-like activity and even "molecular cells" – molecules that were large enough to contain interior compartments in which other molecules might be imprisoned. He was awarded the Nobel prize in 1987 for his innovative approach to organic chemistry.

Cram shared his Nobel with Jean Marie Lehn, who also was interested in mimicking the biochemical world, using abiotic molecules. The basic functions of supramolecular species, in Lehn's view [6], are "molecular recognition, transformation (catalysis), and translocation " – properties characteristic of biomolecules, particularly proteins.

Supramolecular chemistry involves making very high molecular-weight compounds, and some of Cram's creation have a dozen or more benzene rings, each containing six carbons, and associated hydrogens, not to mention the ether linkages. Lehn, likewise, created complicated polycyclic structures that had cavities, clefts, bridges, connections, functional groups, and reactive sites. With these large surfaces, Cram and Lehn were self-consciously trying to mimic the functionality available to biomolecules.

Whilst the original inspiration for supramolecular chemistry was biology, more recent chemists have devoted themselves to producing in miniature a whole range of functional objects found in the macrosphere, such as molecular cages and sponges. These have real world applications in the form of drug delivery systems or possibly toxin removal systems. Other chemists have sought to turn molecules, like crown ethers, into electronic devices. Some delight in creating polymers that self-assemble into daisy chains. Supramolecular chemists and nanotechnologists are hard to tell apart, as many of them share the same toys – nanoparticles like dendrimers, buckyballs and zeolites, objects that will be described more fully in the next chapter.

A delightful book for the technophile, *Molecular Devices and Machines* [8], which was written by three Italian chemists, reports on a whole cornucopia of nanoscale organic molecular mechanisms consciously designed to replicate functions common in macroscale machinery. Among these are molecular machine parts such as rotors, cogwheels, gears, paddle-wheels, ratchets, gyroscopes, brakes, pumps, locks, tweezers, and even harpoons. The illustrations alone are worth the price of the books, depicting side-by-side the chemical structure of a molecule next to the macroscale physical object that it represents. An entire chapter is devoted to the process of threading one molecule through another molecule. Another chapter reports on light-harvesting antennae, mimicking the process of photosynthesis, to convert incoming photons into electric or chemical energy. Several chapters are devoted to molecular electronics, including molecular memory devices and logic gates.

Although supramolecular chemists build very large complicated molecules that interact with each other in mechanical ways, similar to what Drexler and molecular nanotechnologists have predicted, they manufacture their creations the old-fashioned way, through chemistry. Although they may resort to extensive automation, there is nothing yet like Drexler's molecular assembler available yet to snap molecules together like Lego blocks.

References

1 Feynman, R. (**1960**) There's Plenty of Room at the Bottom. *Engineering and Science* 20: 22 (February), **1960**. Reprinted in *Miniaturization*, edited by H. Gilbert, Reinhold, **1961**.

2 Bawa, R., Bawa, S.R., Stephen, T., Maebius, B., Flynn, E., Chiming Wei. *Protecting new ideas and inventions in nanomedicine with patents.* Nanomedicine, in press.

3 Schrödinger, E. *What Is Life?* Cambridge: Cambridge University Press (**1944**).

4 Stent, G., That was the molecular biology was. *Science* (**1968**). 160: 390–395.

5 Fischer, E., *Ber.* 39: 530 (**1906**). Translation taken from Greenstein, J.P. and Winitz, M., *Chemistry of the Amino Acids*, Vol. 2, John Wiley, 1961, p. 1816b.

6 Cram, D.J. (**1987**) Nobel Address, Chemistry.

7 Lehn, J.-M. (**1987**) Nobel Address, Chemistry.

8 Balzani, V., Venturi, M. and Credi, A. *Molecular Devices and Machines: A Journey into the Nanoworld*, Wiley-VCH. Weinheim, Germany **2003**

Chapter 4
Nanotools

In the California gold rush of 1849, the hardware stores and whorehouses made more money than most of the miners. Likewise, the nanotech gold rush needs picks and shovels. Intimate services are beyond the scope of the present volume.

While it may be a while before some of the more esoteric nanoscale devices reach the market, the market for nanotools is here and now. Generous funding through the National Nanotech Initiative has created demand for such devices as the electron microscope, and the various derivatives of the scanning probe microscope. Another category of interest is the nanomanipulators – just exactly how do you pick up and place a nanoparticle, or eventually, even a single molecule?

The Electron Microscope

The electron microscope (EM) is one way of characterizing materials at the nanoscale. As previously noted, the EM was conceived in the 1930s, so why should we regard it now, in the 21st century, as an exciting tool for the Nano Age?

Most of us have encountered an optical (light) microscope in high-school biology class, if nowhere else. Replace the light source with an electron source, and you have the basic idea for an electron microscope. The "electron gun" in an electron microscope is very much like the electron source in the cathode ray tube that used to be the universal monitor for television sets.

A sample is illuminated by a focused beam of electrons and the resulting image is amplified in size through a series of lenses, allowing us to see what would otherwise be invisible. Of course, our eyes cannot see electrons, so our examination of the resulting image is necessarily indirect. The electrons are focused on a fluorescent or phosphorescent screen, similar to a television screen or a computer monitor (Fig. 20).

So what is the advantage of electrons over light waves? The resolution that can be achieved in microscopy is limited by wavelength – the smaller the wavelength, the smaller the size of objects that can be imaged. The resolution of an electron microscope is 100 000 times better than the best optical microscope. In a good light microscope at high amplification, human cells can be seen plainly, and large

The Nanotech Pioneers. Steven A. Edwards

ISBN: 3-527-31290-0

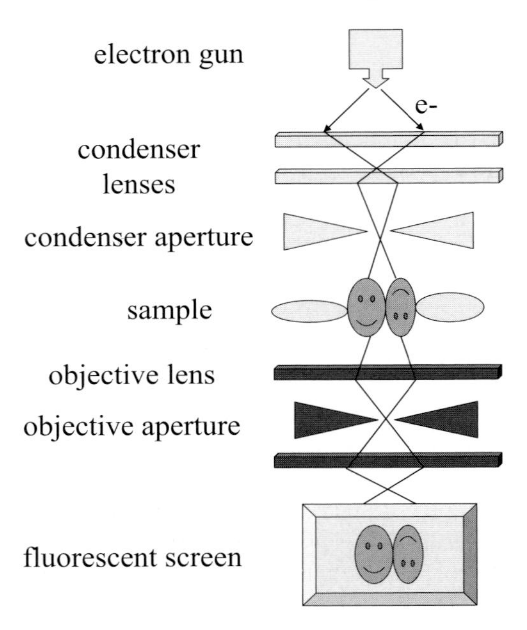

Figure 20 Schematic diagram of a generic electron microscope. Adapted from a figure by Robert Emery, St. Francis Xavier's College, NSW, Australia.

organelles such as the nucleus are also visible. Some bacteria can also be seen, but only as tiny dots. Viruses are beyond the limit of resolution. The images we have seen of viruses are created with electron microscopes.

The wavelength of an electron is small enough that it should be possible to see image individual atoms using electron beams, according to quantum physics. When Ruska received the Nobel Prize in 1986, individual atoms had in fact been imaged – but not by the electron microscope. Instead, the first images of individual atoms were created by the scanning probe microscopes invented by his co-laureates Heinrich Rohrer and Gerd Binnig.

A general problem with scanning probe microscopes is that they are limited to probing the surface of a sample. A transmission EM, on the other hand, can see right through small samples and can focus at any level. The problem with the EM has been that, as a practical matter, they just could not be focused all that well. Until recently, the EM was somewhat myopic. In his lecture, "There's Plenty of Room at the Bottom," Richard Feynman challenged physicists to improve the electron microscope so that it could perform up to its potential. It took 70 years, but with a little help from computer scientists, the physicists and engineers have finally got the EM working correctly.

In 2004, Steven Pennycook, at Oak Ridge National Laboratories, set the record for resolution with a scanning transmission EM at 0.6 Angstrom (an Angstrom is one-tenth of a nanometer), using a computational-intensive technology called aberration correction, which might be thought of as an automatic focus for the EM (Fig. 21). As Pennycook puts it, aberration correction gives you "...the ability to focus 50 different lenses simultaneously." For ordinary mortals this is impossible, but it's not too much of a stretch for a computer. At sub-angstrom levels, atoms are clearly visible, looking very much like the little billiard-ball models in your chemistry book.

Pennycook came to Oak Ridge from Cambridge University, and the very same Cavendish laboratory where the electron was discovered in 1897. He retains his British accent and that air of detached bemusement cultivated by Cambridge professors.

Sitting at his desk, Pennycook pulled out a stack of photomicrographs taken by his wonderful machine and pleasantly answered my uninformed questions. My first one was an old one: "What good would it be to see individual atoms distinctly?" asked physicist Richard Feynman, in his famous lecture, "There's Plenty

Figure 21 Stephen Pennycook, of Oak Ridge National Laboratory, with record setting electron microscope, which took the resolution down to 0.6 Angstrom, less than the diameter of a hydrogen atom. Photo by author.

of Room at the Bottom." Feynman answered his own question: "It would be very easy to make an analysis of any complicated chemical substance; all one would have to do would be to look at it and see where the atoms are." Slowly, through Pennycook's explanations, it became clear what the ability to see atoms in EM could mean.

To the electron, an atom is mostly space, and so most of the electrons penetrate right through relatively small samples. Contact with the nucleus of an atom causes the electron to be reflected, just as light is reflected from a solid object. The larger the nucleus, the more electrons are reflected, and the brighter it appears to the observer. What this means, in practice, is that the positions of each kind of atom in a sample can be identified. In a salt crystal, for instance, the sodium atoms would appear almost twice as bright as the chloride atoms. Feynman's dream has become a reality.

There's more. Pennycook showed me how he could focus down through one atomic layer to the next. So it should be possible to look at slices of a material at the atomic level, in the same way that a medical computed tomography (CT) scan looks at slices of a human body. From these, a three-dimensional picture of a sample can be constructed, atom by atom.

Pennycook and his crew frequently look at interesting samples sent to him from semiconductor companies. These companies "dope" semiconductors with impurities to change the conductivity or to change the structural characteristics of the material. So far, this has been an exercise in witchcraft more than science. But with a new, improved EM, they can actually see the structure of the new material they are creating.

As I watched, two of Pennycook's colleagues analyzed the structure of a sample of a semiconductor material they had given. The news, apparently, was not good – the dopant atoms, instead of spreading throughout the material, were congregating in isolation by themselves. But for me it was fascinating to watch as the microscope scanned through the sample and the computer created a real-time image of the material, atom by atom.

The Department of Energy has put together a group to build the next generation transmission electron aberration-corrected microscope (TEAM), which will have a resolution of 0.5 Å, or smaller. The team includes researchers from Argonne National Laboratory, Brookhaven National Laboratory, Lawrence Berkeley National Laboratory, Oak Ridge National Laboratory and Frederick Seitz Materials Research Laboratory. The instrument will be built by FEI Company, an American manufacturer of transmission electron microscopes and other nanotools, and Corrected Electron Optical Systems, a German firm.

The latest commercial product from FEI is the Titan scanning transmission electron microscope, which almost matches Pennycook's record for resolution. Named after the giant demigods which once walked the Earth, the machine is 4 meters tall, weighs 2000 kg, and uses up to 300 kV. The Titan has a resolution down to 0.67 Å. Orders for the Titan, which will be shipped starting in 2006, have ballooned FEI's backlog to record levels

Scanning Probe Microscopes: STM, AFM and Variants Thereof

The direct "visualization" of individual atoms first became possible in 1982 with the invention of the scanning tunneling microscope (STM). This won its inventors, IBM Zurich researchers Gerd Binnig and Heinrich Rohrer, the 1986 Nobel Prize. Because the STM depends on a current passing from sample to tip, only conducting samples could be imaged.

The components of an STM include a voltage source, probe tip, a piezoelectric material (which changes shape in response to current) to control tip movement in three dimensions, a voltage meter, and sufficient computing power both to control tip movement in real time. The scanning tip is very sharp, ideally only a few molecules at the tip, and is usually made of platinum/iridium or tungsten (Fig. 22).

Imaging with the STM involves moving a tip over a surface to obtain topographic information about the surface; this is similar conceptually to the way in which a blind person reads Braille. The STM does not physically "feel" things, however: it relies on the "electron tunneling", a quantum-mechanical phenomenon that is manifested by a current induced by the voltage differential between the scanning tip and the sample. The level of the tunneling current is directly proportional to the distance between the tip and the surface. The closer the tip is to the surface, the higher the current.

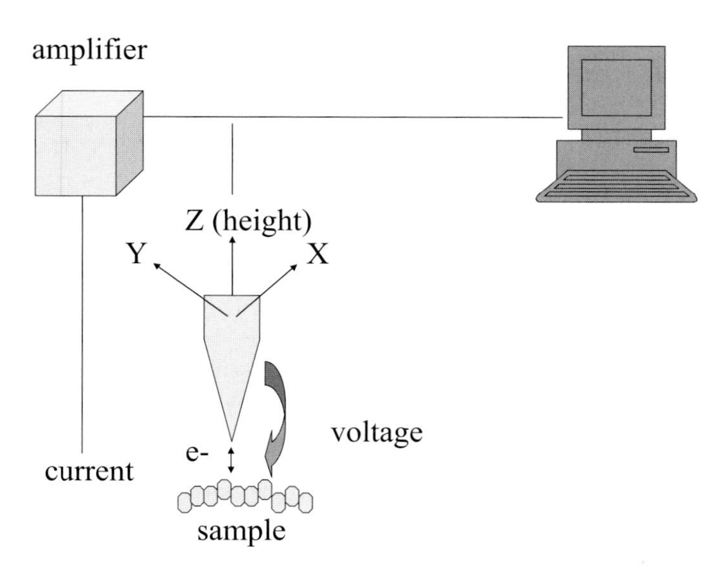

Figure 22 Simplified diagram of a scanning tunneling microscope. A very fine probe is moved over the surface of the material under study, and a voltage is applied between the probe and the surface. Electrons will "tunnel" or jump from the probe to the surface or vice-versa, depending on the polarity of the voltage. This weak electric current is exponentially dependent on the distance between the probe and the surface. Tip movement is controlled in three dimensions using piezoelectronics.

IBM now has a project in progress called Millipede, which would use the STM concept for computer memory. Essentially, Millipede consists of thousands of STM tips that would read the presence or absence of an atom as the 1s and 0s of computer memory. The STM would also be used to change the position of the atoms on the surface, and thus the memory's content. Of course, any surface is also created of atoms, but if you will recall Don Eigler's IBM logo (see Fig. 18), the larger xenon atoms on the nickel surface appeared to the STM like so many peach pits on a plate.

The STM is limited in that it can only be used on a sample that can conduct at least a small current. The atomic force microscope (AFM), which Binnig, Christoph Gerber and Calvin Quate developed in 1986, avoids this limitation by measuring the tiny deflections that a sharp probe experiences when dragged over a surface

The most general scanning probe instrument, the AFM resembles nothing so much as a turntable used to play old-fashioned vinyl records. It doesn't "see" so much as "feel" its way over the surface of a sample. As it scans across a surface, the AFM probe, which is mounted on a cantilever is deflected up or down, and this deflection can be measured, usually through the use of a laser.

What we see as a flat smooth surface can look more like a jeep trail to an AFM tip. However, the deflections up and down of the AFM tip are a little different than what we experience when driving over bumps and potholes. At the atomic scale, there is a repulsive force between two atoms that are adjacent to one another but not covalently bound. The AFM uses a probe that is at most a few atoms in width. It is the repulsive force between atoms in the sample and atoms in the AFM tip that constitutes the "atomic force" which causes the tip to move up and down. Atoms are not really like little billiard balls with hard shiny surfaces; they consist of a relatively stable nucleus orbited by clouds of electrons. Innumerable scans over a surface allow a computer tracking the tip to build up a three-dimensional picture of the surface over which the AFM travels.

Though IBM researchers invented the STM and AFM, Big Blue has not tried to commercialize its inventions except by licensing its patents. The main players in the STM and AFM markets have been Digital Instruments, now a division of Veeco, Omicron Nanotechnology, RHK Technology, Park Scientific, TopoMatrix (also folded into Veeco) and Molecular Imaging, which is now part of Roper Industries. A relative newcomer, Berlin-based JPK Instruments AG, has been successful by focusing primarily on the life sciences market. Pacific Nanotechnologies has also entered the market recently. Overall, Veeco Instruments, through its acquisitions, is by far the dominant market player. However, a few of Veeco's researchers have moved down the street in Santa Barbara to establish Asylum Research, which concentrates on the life science market.

So far, the major uses of AFM and STM have been in metrology and quality control in the semiconductor industry. It has only been during the past few years that the life science market for these instruments has really been considered. AFM has found uses in measuring the force required to unfold proteins and has also been used to visualize the surfaces of viruses.

Another use of the AFM is dip pen nanolithography, which was described briefly in Chapter 3. Essentially, this technology uses the AFM tip as a nanoscale fountain pin to trace out patterns on a surface using molecular "inks". The greatest use of dip pen nanolithography is expected to be for drawing nanoscale electronic circuits, although the technique is also be valuable in biomedical research.

There are a number of variants on AFM that have come into use during the past few years. One of these is the near-field scanning optical microscope (NSOM). Although resolution of the optical microscope has always been limited by the wavelength of light, it turns out that there is a way around this limitation. By mounting the light detector (usually an optical fiber) on an AFM tip, at a position that is less than a full wavelength away from the sample, it is possible to increase resolution of the light microscope ten-fold. Essentially, the stream of photons is being captured before they diffuse into a wave.

An analogy that is often used to explain NSOM is that of a stethoscope. The wavelength of sound is on the order of meters. If you stood a couple of meters away with a very sensitive stethoscope, you could tell that a beating heart was in the patient, but that's about all. By placing the stethoscope on the chest you can locate the heart very precisely.

Another variant of the AFM measures Van der Waals forces instead of atomic repulsion. By raising the tip to about 2 to 20 nm from the surface of the sample, one can measure the deflection that occurs because of the attraction of the tip to the sample. Other AFM spin-offs measure chemical reactivity, heat capacity, or temperature changes across the surface of a sample

AngstroVision

AngstroVision is a small nanotech company that is creating a new type of imaging device which will allow the examination of nanoscale objects in real time. According to the company's co-founder, Scott Mize (also the current president of Foresight Institute), molecular vibration or Brownian motion presents serious challenges for imaging devices. Scanning techniques, which acquire images by sensing a surface point by point, require a relatively long time to construct a whole image, with some minimum amount of time between the acquisition of two successive points. Since the speed of image acquisition is less than the speed of the molecular motion, the result obtained is some type of statistical average of the surface you are trying to measure. Paradoxically, the higher the resolution of the technique becomes, the less accurate is the description, because the molecular movement occupies a greater percentage of the range of resolution. For instance, using an electron microscope, the nuclei of atoms appear to nearly as large as the atoms themselves should be, which is clearly an artifact of the technique. An accurate measurement of surfaces at the nanoscale and beyond, says Mize, can only be done when the speed of image acquisition is much greater than the speed of molecular displacement.

Another problem is interaction between the imaging tool and the object being imaged. Any mechanical or electrical interaction has the potential to alter the material being sensed, and this is hard to control and account for these interactions with devices such as an STM or AFM.

AngstroVision's technology seeks to overcome these limitations by using light interferometry as a measurement technique. Three-dimensional images would be created by the interference created by the collision of visible light beams, somewhat in the way that holographic images are created, only at a sub-nanometer scale. AngstroVision, however, has released little in the way of information about how this would work, in practice. They are apparently afraid of releasing details ahead of the issuance of patents.

The images would be created at the speed of light in space. Nothing, according to Einstein, moves faster than light, so this mode of image should keep ahead of molecular motion. The speed of image capture is limited only by the speed of the photon sensor used, such as a CCD device, like those in a digital camera. This result is reportedly metrologically accurate three-dimensional images of the topography of surfaces acquired very rapidly at resolutions in the single nanometers. Moreover, the technique is non-destructive and can be used in a wide range of environments, including the ambient environment and in solution.

AngstroVision's initial market focus will be on nanomaterials characterization and quality assurance. As the imaging technique is non-destructive, AngstroVision sees significant biological applications in the biosciences, such as the analysis of cells, drug discovery, and diagnostics. According to Mize, AngstroVision's device will allow scientists "... to look into the nanobiological world in its natural state in real time. This has never been done before. We anticipate that they will be able to see dynamic behavior and interaction on the molecular level, which is fundamental information for understanding the mechanics of life itself."

AngstroVision expects to ship its first products in 2005.

Nanomanipulators

Say you've got a nifty nanoscale object that you want to look at, test or just play with. The EM or the AFM give you a way to see it, at least indirectly, but what if you wanted to turn it over, and check out the other side? Or what if you wanted to pull on it, to see how much force it would take to break it? This is not something you can do with your own fat fingers or even a fine pair of tweezers. You need a special tool called a nanomanipulator.

Fortunately, there are small companies who have anticipated your need. Zyvex, mentioned previously, is one of these. Their S100 nanomanipulator consists of four tungsten needles beveled to very fine points. These needles are mounted such that they can be moved in three dimensions. They converge on the sample from four directions, with coarse and fine controls providing you with the ability to pin down your sample or to clamp it between needles. Movements can be controlled with a fine resolution of less than 5 nm. This is done using joysticks, just like a videogame, and the nanomanipulator uses a PC interface. The needles can be outfitted with NanoEffector probes; these are tungsten wires only 50 nm across at their tips that allow the probing of electrical circuits and carbon nanotubes.

3rd Tech's Nanomanipulator SPM Visualization and Control System, an interactive interface for a scanning probe microscope (SPM), was developed at the University of North Carolina. Force feedback and three-dimensional visualization allows you to have a tactile sense as the instrument is steered through nanospace. It also provides record and replay functions, and it serves as an interface for Thermomicroscope's SPM.

The Nanomanipulator uses what 3rd Tech calls the SensAble Technologies Phantom Desktop, a haptic display device that provides continual force feedback for investigators; so when you prod something with your nanomanipulator it "feels" hard, or spongy, or whatever. While working with the nanomanipulator, an automatic lab notebook keeps track of session data, including manipulation data and parameter values.

The Nanomanipulator has already been used extensively on life science projects, including determining the rupture strength of DNA and adenovirus capsids, the differences between normal and hemophiliac blood clot fibers, and the strength of microtubules (a cellular structural protein).

An actual set of molecular tweezers derived from azobenzene is described in Chapter 12.3 of *Molecular Devices and Machines* [1]. The tweezer-like action depends on a photo-induced change in chemical structure, the flipping (*trans-cis* isomerization) of azobenzene groups. The tweezers are capable of grabbing large metal ions. Several other possible photoinduced tweezers are also suggested in the chapter, but these chemical tweezers do not yet appear to be under commercial development.

Reference

1 Balzani, V., Venturi, M. and Credi, A. *Molecular Devices and Machines*, Wiley-VCH (**2003**).

Chapter 5
Nanoparticles and Other Nanomaterials

While nanotech may eventually encompass such science fiction staples as nanoscale robots and memory implants, the nanoscience that is being commercialized already is concerned mainly with new materials (Table 5).

Thin-layer technology has been used for some time in the semiconductor and electronics industries. Newer uses include the polymer light-emitting diodes that are used in electronic displays. We will cover these in greater detail in Chapter 7, on nanoelectronics.

Nanoencapsulation is being used for the delivery of drugs that would otherwise be insoluble. Several examples will be given in Chapter 8, on nanobiomedicine.

Zeolites are naturally occurring minerals formed from silica and aluminum that feature channeled crystals with a very large surface area. Macroscale zeolite crystals are used in water softeners and in water purification.

Zeolite nanoparticles are employed as catalysts in the petrochemical industry and for environmental remediation. Zeolite surface chemistry speeds up certain chemical reactions. Because zeolites are nanoscale particles, they remain in solution, whereas larger particles would settle out. Literally thousands of different zeolite products have been created in an industry that dates back to the 1950s.

Table 5 Nanoparticles and nanomaterials under development.

Atomic clusters	Quantum wells, quantum dots
Thin layers (100 nm or smaller)	Luminescent polymers, coatings, self-assembling monolayers
Fibers	Carbon nanotubes, nanowires, "smart fabrics", nanocomposites
Nanocatalysts	Zeolites
Nanocapsules	Nanomicelles, block polymer capsules, protein capsules
Nanoparticles	Dendrimers, fullerenes

The Nanotech Pioneers. Steven A. Edwards

ISBN: 3-527-31290-0

In this chapter, we will concentrate on more recently developed nanoparticles: the buckyball, the carbon nanotube, quantum dots, and dendrimers. Lux Research and the intellectual property firm Foley and Lardner found that over 200 U.S. patents were issued for these four categories, as shown in Figure 23. Before 1990, there were no patents issued on buckyballs, nanotubes or quantum dots, and only a handful on dendrimers. The last item mentioned in the figure, nanowires, will be discussed in Chapter 7, on nanoelectronics. Much of the intellectual property regarding nanowires has been locked up by a single, small entrepreneurial firm called Nanosys, located in the San Francisco Bay area.

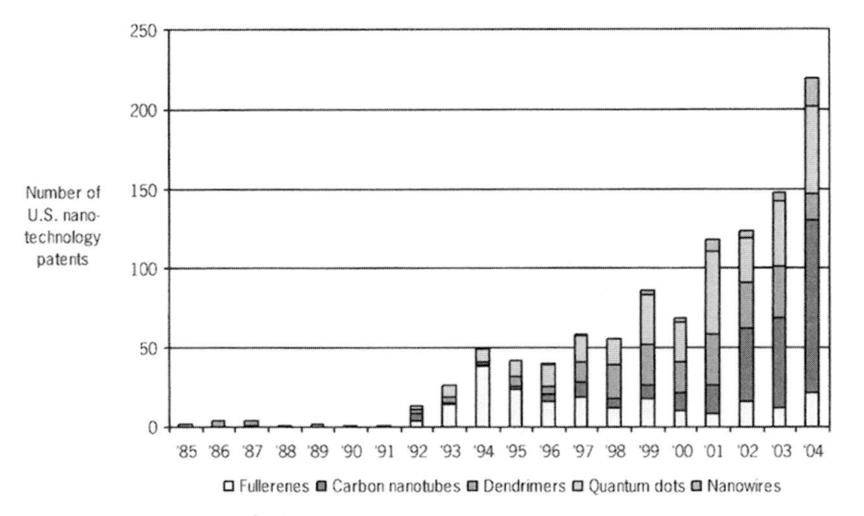

Figure 23 Patents issued for fullerenes, carbon nanotubes, dendrimers, quantum dots, and nanowires. Source: "The Nanotech Intellectual Property Landscape," March 2005, Lux Research and Foley & Lardner LLP.

Discovering the Buckyball

Buckminsterfullerene (also known as buckyball or C60), a sixty-carbon molecule shaped like a soccer ball (Fig. 24), was discovered, named, and its structure deciphered over ten days in September 1985, by five scientists at Rice University. Richard Smalley, Harry Kroto, and Robert Curl were awarded the Nobel Prize for the discovery of the buckyball in 1996 (unfortunately for Sean O'Brien and James Heath, the rules only allow the prize to be split a maximum of three ways). Kroto also had the peculiarly British honor of being awarded a knighthood.

The buckyball discovery was pure serendipity, a case of luck favoring the prepared mind. Kroto compared it to winning the lottery.

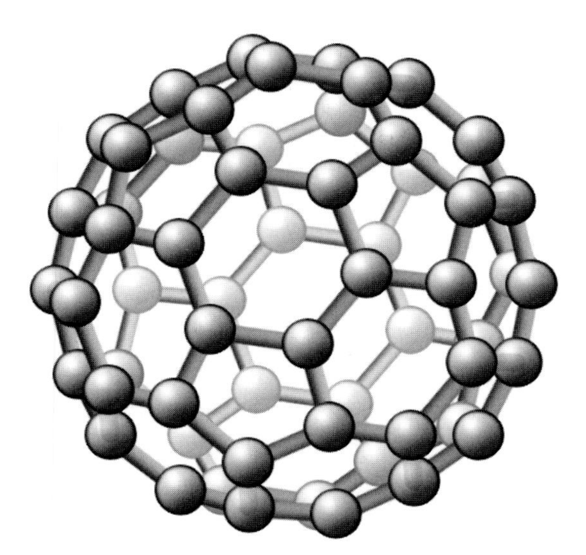

Figure 24 Buckminsterfullerene C60. Contains 60 atoms of carbon, each of which is bound to 3 other carbons in an alternating arrangement of pentagons and hexagons. Reproduced Courtesy of Samantha Shanley, University of Bristol.

As it turns out, the Rice team was not actually the first to observe evidence of buckyballs. A spectroscopic band corresponding to C60 had appeared earlier in a report published by Exxon scientists, but their interest was in finding new catalysts and they failed to appreciate the significance of the large carbon molecule. A Japanese scientist, Eji Osawa, is credited with first predicting in 1970 the sixty-carbon buckminsterfullerene structure, although he did not call it that. The experimental confirmation was due to a fortuitous combination of scientists in the right place at the right time, good scientific intuition, and not incidentally, having the appropriate instruments.

Sir Harry Kroto is a man of insatiable curiosity. For example, when he visits a new city, he says, he just gets lost in it, following roads at random, whichever way seems most interesting. His approach to his scientific career is much the same. Perhaps if he were an American chemist, he might have a more pragmatic bent, but as a Briton he was more than willing to look in directions that seem to have no obvious practical application. One of these directions was out into interstellar space. Kroto and his students had been engaged in making nitrogen-containing long-chain carbon molecules, originally for the purpose of examining the way that they bend and rotate. At around the same time, it had been discovered that carbon molecules were floating around in space – in fact, the new science of astrochemistry had discovered a whole prebiotic primal soup of organic molecules floating around the center of the galaxy, as if waiting for the chance to come together to create life. Since he could recognize the spectroscopic signature of his newly created molecules, Kroto wanted to know whether his new creations might have preexisted way out there. Kroto was able to convince one of his former colleagues to search for his long-chain carbons, and sure enough, they were there, hanging out within an interstellar cloud around Taurus constellation, among other places. In fact, they were there in much greater quantities than one would have expected, based on existing theory. In time, an explanation presented itself – cool red giant carbon stars might be pumping out Kroto's pet molecules into space.

In 1984, Kroto visited Richard Smalley's laboratory in Houston and was introduced to a new piece of equipment, the laser-vaporization supersonic cluster beam, a Tom Swift–Electronic Wizard type of apparatus that you aren't likely to find in your college O-Chem lab. It uses the power of a pulsed laser beam to vaporize material from a metallic disc. The vapor is swept up in a cloud of helium, which rapidly cools the material, which then begins to cluster. The material is then sprayed through a nozzle into a vacuum, cooling it further, and the results are analyzed with a mass spectrometer, an instrument that separates molecules according to their molecular weight.

Kroto decided that Smalley's instrument was about as close as you could come to duplicating the action of red giant carbon stars here on Earth. He wanted to see if his long-chain carbons might form *de novo* after blasting a graphite disk with a laser. When experiments were begun in September of 1985, the investigators immediately found the 5- to 9-chain carbons that Kroto was looking for, along with much larger carbon molecules. In particular, there was an annoying band which contained 60 carbons and always appeared much larger than the others. It was not the peak itself that was troubling – there were many peaks to choose from – it was because the peak was so disproportionate. Under optimized conditions, almost nothing but C60 was produced. What was magic about the number 60 that made it a favored form for carbon molecules to assume? It was not an easy problem to solve: how many different ways are there to stick 60 carbons together? Each carbon can bind to from two to four neighboring carbons. Just think about it for a while! The first clue was that the molecule was only carbon – a carbon molecule with ends or edges one would expect to include hydrogen. Somehow, the reactive ends had disappeared.

Model building began in earnest – not as you might expect, with powerful computers seeking a best fit – but with toothpicks and jelly-beans, and later with paper cut-outs of hexagons and pentagons. This is precisely the sort of high-tech molecular modeling that Watson and Crick had used to crack the structure of DNA some forty years earlier. The solution to the C60 structure finally presented itself after Kroto, once a student of design, remembered his visit to one of Buckminster Fuller's architectural creations, a geodesic dome at Montreal's Expo '67. C60 resembled two Fuller domes stuck together to form a sphere – hence the name, buckminsterfullerene, or more informally, buckyball.

The most startling thing about a buckyball is its absolute symmetry. Every carbon is bound to three others, and every carbon is at the vertex of one pentagon and two hexagons. Since every carbon is equivalent, the nuclear magnetic resonance spectrum of C60 reveals a single line. Buckminsterfullerene is a beautiful thing.

Its very "thingness" seems to put the buckyball in a different category than most compounds. Chemical structures, even to the chemist, have an abstract quality. We draw out the structures of chemicals as a way of thinking about them, a pedagogical tool. But we don't really expect them to look the way they are drawn; indeed, we may have multiple representations of a given structure, depending on what point we are trying to emphasize. Buckyballs, with their soccer ball shape, however, seem more like an object then a chemical. And now it is possible to image these creatures with scanning probe microscopes. Sure enough, they look just like you would expect, like little soccer balls.

Once a thing is identified and named, it becomes visible, as if the identification caused it to come into being – suddenly you find it everywhere you look. Though it took a laser vaporization supersonic cluster beam to first identify buckminsterfullerene, it was soon found as a byproduct of soot formation. Scrape the inside of your chimney and you will likely get a few buckyballs on your fingers.

Buckminsterfullerene is so far only a wonderful example of carbon chemistry. But what can you do with it? One group of enterprising researchers assembled a group of buckyballs on a flat surface into the shape of an abacus. Although in principle you could use such an abacus to do calculations, it is not in any way practical, though it makes a great piece of tiny art-work, assuming you have an atomic force microscope to view it with.

Table 6 What can you do with a buckyball?

Company	Application
C-Sixty/Merck	Anti-oxidant pharmaceuticals, other drugs
Luna NanoWorks (trimetaspheres)	MRI contrast agents, light-emitting diodes
Others	Rocket fuel, lubrication

One small New York-based company called C-Sixty hopes to use buckyballs as the basis for new drugs. Surprisingly, unmodified buckminsterfullerene has proved to be an effective inhibitor of the HIV protease. Coincidentally, it happens to fit very nicely into the active site of that enzyme. The company also hopes to use the buckyball as a kind of spherical three-dimensional co-ordinate system, which they can substitute as necessary to make drugs with the correct structure to interact with target proteins. Finally, they have discovered that buckyballs are superb anti-oxidants (Table 6).

Degenerative diseases and indeed even ordinary aging processes are caused, in part, by intracellular oxygen free radicals – reactive molecules with an unpaired electron. C-60 fullerene has thirty carbon–carbon double bonds, all of which can react with a radical species. This should given them unusual power to halt the progression of diseases caused by excess free radical production. The fullerene forms a bond with a radical very quickly essentially every time it encounters one. Serendipitously, C-60 also has the property of accumulating within the mitochondria of the cell where most superoxide radicals are created.

C-Sixty has a collaborative project underway with Merck to use modified buckyball molecules as treatments for the oxidative stress that occurs in cardiovascular disease, strokes, and neurodegenerative diseases, such as Alzheimer's.

Buckyballs are hollow cages. There is a 0.4 nm cavity within each one, in which in theory it is possible to enclose any element of the periodic table. Chemists are unable to resist such a challenge, and by now they have crammed most of the available elements into buckyballs for the sheer pleasure of it. However, some of these creations may also have valuable attributes, and a small entrepreneurial company called Luna Nanoworks is hoping to develop commercially attractive buckyball variants.

Luna NanoWorks is located in Danville, VA, once a center of the tobacco industry. The city fathers, in order to fight rising unemployment, have been promoting the area's charms to the nanotech industry, with much help from the state government. Luna has taken over a former tobacco warehouse on the banks of the Dan River, and has gutted the inside, leaving only beams and the foot-thick brick walls. In this unlikely space they are building what they hope to be a nanotech powerhouse, capable of producing ton quantities of nanomaterials. An old railroad spur backs up to the warehouse/factory, awaiting the day when this promise is fulfilled. A cobblestone street in front of the building completes the picture of a nineteenth century setting for this high-tech upstart.

The president of Luna NanoWorks, Stephen Wilson, is a former chemistry professor from New York State University, and is also the scientific founder of C-Sixty. Wilson was one of the first to see the possibilities inherent in buckyballs – cornering on behalf of C-Sixty much of the then-available intellectual property surrounding medical applications of that curious molecule. A devotee of fine wine and exotic food, he is one of those rare scientists who enjoys the finer things in life when he leaves the lab. The move to Danville has allowed he and his wife Susan to move out of a New York apartment and into a fine old Virginia antique of a house.

In addition to the commercial production carbon nanotubes and buckyballs, the company hopes to commercialize what it calls trimetaspheres (Fig. 25). These are a larger version of the buckyball, with 80 carbons caging up to three metal or rare earth atoms, such as scandium, lanthanum or yttrium, which are covalently bound to nitrogen. The nitrogen complex spins freely within the larger cage of carbons. Trimetaspheres were invented by accident when an air leak contaminated a reactor being used to make buckyballs. Rather than merely discard the preparation and start over, Harry Dorn, a professor at Virginia Tech University decided to characterize the anomalous chemical species that were thus formed. Science was once again served by the unplanned experiment and a curious mind.

According to Wilson, trimetaspheres have potential uses as contrast agents for medical magnetic resonance imaging, to make light-emitting diodes, and potentially for molecular electronics and computing.

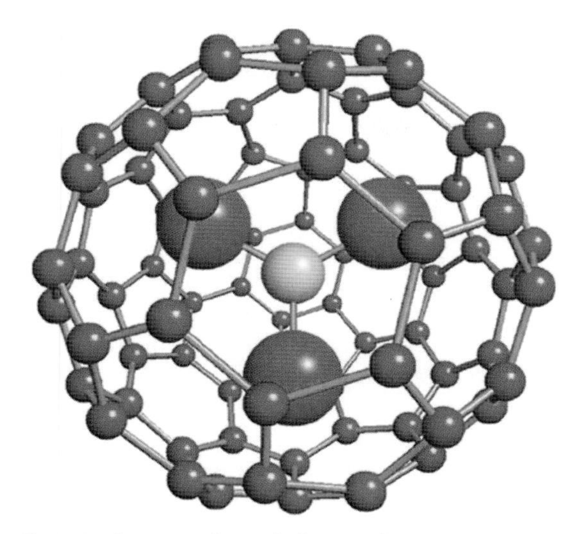

Figure 25 A trimetasphere. The brown spheres represent carbon atoms. The larger blue spheres are metal or rare earth atoms. The yellow sphere is a nitrogen atom. The nitrogen complex spins freely within the carbon cage. Reproduced with permission from LunaNanoworks, Danville, VA.

Other applications of buckyballs that have been suggested include their use as lubricants and even as rocket fuel. The lubricant idea makes sense, as another carbon compound, graphite is already widely used that way. Buckyballs would, in theory, be even better. Imagine surfaces covered with slick little ball bearings only 1 nm in diameter. However, the current cost of production is prohibitive and probably will remain so for the foreseeable future. It is also not clear that putting large volumes of buckyballs into the environment would be safe (see Chapter 11).

In addition to the C60 buckyballs, larger variants have been created with 70 or 80 carbons, with or without caged molecules. Researchers from Xiamen University and the Chinese Academy were able to construct a smaller version of the buckyball, one that only has 50 molecules. Chemists think that fullerenes smaller than C60 may have odd electronic and magnetic properties because of their shape, which is intermediate between a sphere and a disk.

Bernard Erlanger, an organic chemist and microbiologist from Columbia University, was the first to succeed in making a monoclonal antibody to buckminsterfullerene. "Whenever I brought up the subject with friends at lunch, they would ask me why I wanted to do THAT!" remembers Erlanger. "Of course, I could not give them a satisfactory answer." A pure scientist, Erlanger was just curious to see if it was possible. In principle, the buckyball would seem to be the least antigenic molecule conceivable, as it consists of only carbon, the basic unit of all organic molecules from which living things are constructed. As it turned out, with help from lymphocyte cultures, Erlanger was able to make the antibody. Despite being a monoclonal antibody – a single, completely defined protein molecule – Erlanger's antibody cross-reacted with carbon nanotubes (see below) which, like buckyballs, are composed simply of carbon bonded to carbon. Antibodies could turn out to be a way of linking fullerene-based nanodevices to biomolecules for the creation of specific sensors or hybrid bio-nano machines.

Carbon Nanotubes

In 1991, Sujio Iijima, a Japanese researcher working for NEC, was exploring a new way to make fullerenes using arc evaporation with graphite electrodes, when he discovered a whole new class of fullerenes called nanotubes [1]. These are long, cylindrical tubes that are usually capped. Although the nanotubes that Iijima discovered were "multi-walled" – two or more cylinders nested one within the other, like Russian dolls – single-walled nanotubes can now be reliably synthesized. Each carbon nanotube is a single molecule composed entirely of carbon; it may be thought of as a graphene sheet (from graphite, see Chapter 1) with its hexagonal ring structure rolled back on itself to form a cylinder, like a nanoscale bundle of chicken-wire. Figure 26 shows an the image of a nanotube made by a scanning tunneling microscope. Notice the regular twist to the pattern of carbon atoms.

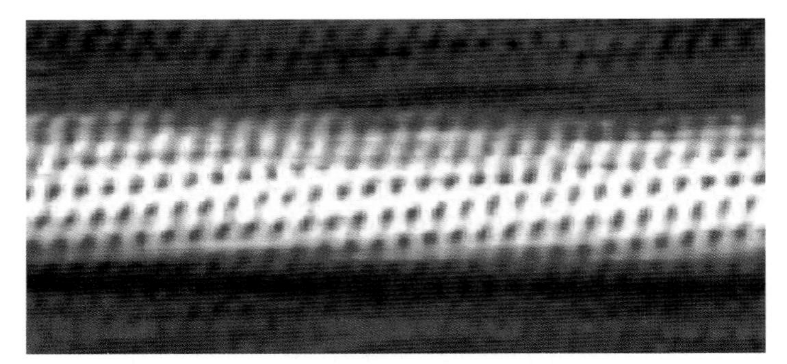

Figure 26 Scanning tunneling microscope image of a single-walled carbon nanotube. The reddish blobs represent individual carbon atoms. Courtesy of Liesbeth Venema, Jeroen Wildoer, and Cees Dekker at the DIMES institute at Delft University of Technology.

Like DNA, carbon nanotubes have very small diameters in relation to their length. The diameter of a single-walled carbon nanotube can be very small (0.4 nm is the theoretical limit), whereas the length of the cylinders is indefinite and may be macroscopic. The record as of now is several centimeters. Similarly, although double-stranded DNA is only 2.0 nm in diameter, its length can extend to hundreds of centimeters. Unlike DNA, which can be spooled into very small spaces (you have several meters worth in every one of your cells), carbon nanotubes are relatively stiff. Nanotubes are also very sharp, and penetrate human cells with ease.

Carbon nanotubes have a host of unusual properties. They are as hard and as heat-conductive as diamonds, and are many times stronger than steel at one-sixth the density. Multi-walled nanotubes have been created that have a measured strength 375 times greater than steel (150 gigapascals versus 0.4). Unlike steel, nanotubes are flexible and elastic, bouncing back to their original shape after being bent or stretched. Nanotubes are already being used in polymer composites that are stronger and lighter than previous carbon fiber materials. Unlike carbon fibers, however, single-wall nanotubes are extremely flexible; they can be twisted, flattened and bent without breaking. They can be even be formed into small circles or bent around corners.

Though Iijima discovered carbon nanotubes quite by accident, they are now an active, quite purposeful area of research. Many companies, both young entrepreneurial companies like Luna Nanoworks, and giants, like the Japanese giant Mitsui, would like to know how to make them in quantity.

In October, 2004, I traveled to Oak Ridge National Laboratories where David Geohegan and his colleagues produce some of the purest (99.9 %) single-walled nanotubes available to date to find out how this trick is accomplished. Good journalist that I am, I had brought a camera to document the event, and so my first action was to try to photograph Dr. Geohegan in his office. Unfortunately, the camera was a digital model that I purchased for my wife and while she can do marvels with it, I have never figured the thing out. Geohegan perceived my distress, grabbed the thing away from me, and within a couple of seconds had succeeded in taking a picture of me.

Figure 27 David Geohegan, Oak Ridge National Laboratories. Photo by author.

Geohegan is all about photography, it turns out. Behind him, on his computer monitor, movies he has taken of Venice occupy his screensaver. Impressive, multi-color visions are pasted on the walls of the office and above his desk; these look like time-lapse frames of supernova in the process of exploding. Actually, it turns out, they are merely microsecond-by-microsecond false-color shots of plasma plumes erupting after laser ablation. I should have known.

Similar to the way that buckyballs were first formed experimentally, single-walled nanotubes are created by zapping a graphite disc with a laser. This creates a plume of vaporized carbon. In the case of nanotube production, this plume is purposefully contaminated with a metal as a catalyst for nanotube production. As the plume of hot carbon cools, a colloid of carbon and metal forms. The exact chemistry here is not known with certainty. What is clear is that the metal catalyzes the incorporation of a carbon into a growing tube; new carbon is added at the metal interface, while the growing nanotube extends away from the metal. The end of the nanotube has a rounded cap, similar to half a buckyball.

To visualize the growth of a nanotube, think of the type of fireworks known as Snakes. These are cylindrical carbon pellets that are placed on the ground and lit at the upper end. The burning end sends up a puffy snake-like twisting cylinder of ash as it burns. In the case of nanotubes, however, the addition of carbon to the growing tube actually comes as the plume of carbon cools. Geohegan, with his cameras, has documented the growth rate of the nanotubes, which is about 1 μm (one thousand nanometers) per second. The eventual length of the nanotube can be measured in centimeters, while the diameter is as small as 0.5 nm. The ratio of length to diameter can be measured in the millions; we're talking really thin, much too thin to see with an optical microscope, but long enough to be macroscopic. A bundle of nanotubes is easy to see.

Geohegan, serving in his role as host, showed me the lab's "tube furnace" – a glass cylinder about 2.5 meters long. Pointed toward one end is a large, impressive laser generator. Downstream, just inside the cylinder proper, is a sample holder that contains the graphite target. Just beyond that, Geohegan's cameras monitor the plume and the formation of nanotubes. A powerful vacuum pump occupies the distal end. The whole apparatus, while reeking of technology, is obviously experimental. This is a one-off device that geeks have rigged up. If this were intended for the large-scale manufacture of nanotubes, there would be a big chrome enclosure with a proud logo on it. The device would come with a three-ring binder full of instructions in pidgin English and a flat-screen computer monitor. But obviously, we are not there yet.

The commercial production of carbon nanotubes is progressing rapidly, however. For instance, Montreal-based Raymor Industries claims to have a plasma process for manufacturing carbon nanotubes that uses methane as a feedstock. Nearly 100 % of the carbon is converted to nanotubes, leaving only hydrogen as a byproduct. Hydrogen, of course, is expected to be the fuel of the future, so the byproduct is also useful and valuable.

There is another way to make nanotubes, and this is called chemical vapor deposition. In this case, metal catalyst particles are coated onto a solid surface (substrate), and a hot carbon vapor is introduced. The metal catalyzes the incorporation of carbon into nanotubes. Oriented by gravity, however the nanotubes push up from the substrate, suspending the metal catalyst at the tip. Because the nanotubes are growing closely adjacent to one another, they become vertically aligned. Geohegan, with his cameras, has also documented this sort of growth. It tends to stop all at once, like a pile carpet, with all the nanotubes more or less exactly the same length. The reason it stops so suddenly is not at all clear.

Chemical vapor deposition has the advantage that the nanotubes grow uniformly, are vertically aligned, and are more or less the same length. Laser ablation, however, produces, almost always, single-wall nanotubes, whereas chemical vapor deposition produces mainly multi-walled tubes. So your choice of synthesis modality depends upon what you want. Single-walled nanotubes are required to make electronic components of defined properties. Multi-walled nanotubes make may make particularly high-strength fibers, although there is still debate about this. Vertically aligned fibers may be important for the creation of field emission devices (see below). Geohegan's group has created both carbon nanotube fibers and something they call nanotube "paper." A Massachusetts company called NanoLab has also manufactured nanotube paper that they claim is useful for the creation of electrochemical electrodes and for cell culture experiments. $1000 will buy you a circle of such paper that is 14.0 cm (5.6 in) in diameter.

It is widely expected that commercial carbon nanotube production will soon be in the range of millions of kilograms per year, using simple, inexpensive carbon feedstocks such as methane, carbon monoxide or methanol with a nanoscale catalyst particle attached to the end of each growing tube. Quality control will be a challenge; ideally, the need is to create nanotube products that are identical in length, diameter, and electrical type.

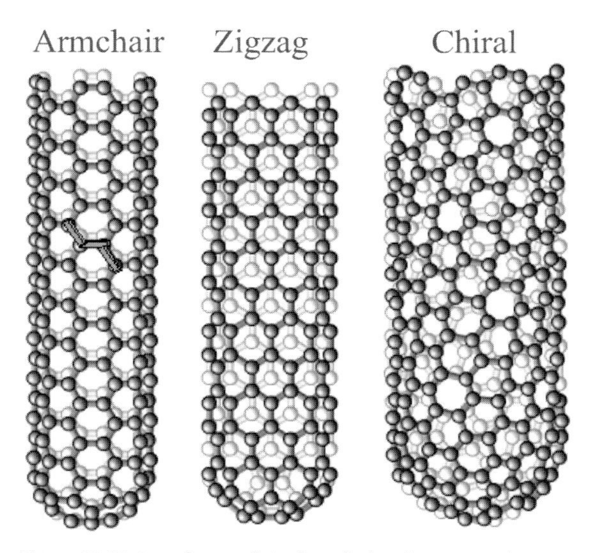

Figure 28 Various forms of single-walled carbon nanotubes.

Nanotubes can be constructed that are either more electrically conductive than copper or act more like the semiconductor silicon. This difference is related to how the graphene sheet is rolled – the carbon hexagons line up slightly differently, depending on the where the vertical axis of symmetry cuts across the chicken-wire pattern. The "armchair structure" (see Fig. 28) is a "metallic" conducting nanotube, whereas "chiral" and "zigzag" nanotubes are semiconducting. Notice how the carbon atoms line up along the vertical axis in the armchair structure. This facilitates electron flow down the length of the tube.

Methods have been developed to separate conducting nanotubes from their semiconducting siblings. It is also possible to convert a semiconducting tube into a conductor by applying a magnetic field. This unusual property leads to the possibility of magnetically controlled switching devices. At very low temperatures, near absolute zero, carbon nanotubes behave like superconductors – that is, they demonstrate essentially no resistance to electron flow.

Table 7 What can you do with carbon nanotubes?

Company or Research Group	Application
Zyvex	Carbon fiber composite, polyurethane additive
Richard Smalley, Rice University	Nanowires, electrical transmission
Nantero	Memory chip component
Cees Dekker, Delft U.	Transistor, logic gates
Nanoproprietary Inc.	Cold cathode for X-ray devices, displays
UC Davis, Mytitek, Inc.	Supercapacitors made from bundled nanotubes on nickel foil
Molecular Nanosystems, GE	Flat panel displays
Samsung	Television displays
Rensallear Polytechnic U., Banares Hindu U.	Filters with nanoscale porosity
Various groups	Biosensors
CSIRO (Australia)	Nanotube fiber "yarn"
Foster-Miller, Inc; Carbon Nanotechnologies, Inc.	Conductive sealants and caulks for aircraft
LiftPort Group, Carbon Designs	Space elevator ribbon

The carbon nanotube's combination of strength, hardness, flexibility, conductivity, and virtual indestructibility has inspired researchers to come up with a host of potential applications for nanotubes, which are in various stages of being realized (Table 7). Some commercial products already exist. Zyvex, for instance, uses nanotubes to spike polyurethane, increasing the wear resistance of this coating. They are also manufacturing a carbon nanotube composite material for use in making bicycle parts.

It is expected that conducting (metal) nanotubes will make excellent nanowires. Richard Smalley, of Rice University, has suggested that nanotube-based wires should be used for electrical transmission. The advantage would be the transmission of electricity over very long distances with much less energy loss than present, due to heating effects. This would allow an electrical grid that could encompass all of North America, so that Mexico could be supplied with hydroelectric power from Canada, if need be (see Chapter 10 on Grand Challenges).

A host of simple devices, mostly experimental, has been created using carbon nanotubes. Charles Lieber from Harvard University, for instance, has invented nanotweezers made from two carbon nanotubes that can be brought together by applying a voltage across them. The tweezers are small enough to pick up individual molecules. Admittedly these tweezers are not very practical in the near term, but they are an example of the nanoscale tools being created.

Another tiny machine, a nanoscale soldering iron, has been built by Alex Zettl, of the University of California, Berkeley. A nanotube sprayed with indium is hooked to a current, which heats up and melts the indium. A nanoscale dollop of indium is driven electronically to the end of the tube. Reversing the polarity of the current drives the indium in the opposite direction. It is not entirely clear why the uncharged indium atoms should be transported electrically. Zettl thinks that perhaps electrons are passing from the indium to the nanotube, leaving the indium atoms with a positive charge.

Researchers in Japan have designed a very cute contraption – a thermometer that is not only smaller than the eye can see, but is also able to measure temperatures hot enough to soften steel. The researchers found that a carbon nanotube filled with the liquid metal gallium can act as a thermometer, according to Yoshio Bando, a director at the Advanced Materials Laboratory of the National Institute for Materials Science (NIMS) in Japan.

The nanotube thermometer is about 75 nm in diameter and 8000 nm in length, which is about 1.5 times the length of a red blood cell. The tiny thermometer can measure temperatures between 50 and 500 °C. This wide range is possible because gallium remains liquid over a broad range of temperatures.

But why would you need a nano-thermometer? The tiny instrument could be used to measure temperature shifts within very small spaces – for example, *in-situ* observation of chemical reaction in a micro-region, or to measure laser effects. Another potential use would be to estimate the thermal effect of an electron beam when a material is observed in a transmission electron microscope (TEM) or a scanning electron microscope

Pulickel Ajayan, a professor of materials science and engineering at the Rensselaer Polytechnic Institute, may have the honor of inventing the most prosaic use yet for nanotubes. Ajayan assembles batches carbon nanotubes onto "handles" of silicon carbide to make the tiniest brushes known to man. These are used for, what else, sweeping up nanoparticles and general tidying up at the nanoscale. Just what every obsessive-compulsive housewife needs! The nanobrushes also make excellent electrical contacts for nanoscale motors and can be used for cleaning out grooves in computer chips. Ajayan confidently expects a commercial market to develop for his brushes as the general trend toward nanoscale devices continues.

Nanotubes are also being used to make electronic components. The first nanotube transistor was demonstrated in 1998 at the Delft University of Technology, although critics were quick to point out that the device did not lend itself to easy manufacturing. Cees Dekker and a group of researchers at Delft University in the Netherlands have combined nanotube transistors to make logic gates, the basic elements used in electronic computing. IBM, Infineon, and NEC are among the companies that take nanotube logic circuits seriously. Eventually, it is thought that nanotubes will make possible extremely fast transistors that operate in the terahertz range, about 1000 times faster than present-day computers.

Nantero, a small entrepreneurial company, is hoping to make what it calls NRAM, nanotube memory chips that ultimately could replace all current forms of electronic memory in computers, digital cameras, cell phones, etc. (see Chapter 7, Nanoelectronics.

Carbon nanotubes also function as "cold cathodes" that emit electrons in response to an applied field. Nanotube cold cathodes have been used to make a commercially available X-ray spectrophotometer, and may also be used as the basis for portable medical X-ray machines. Current X ray devices use metal filaments that must be heated to high temperatures (1500 °C) as electron emitters. Two small entrepreneurial companies, Nanoproprietary, Inc. and Applied Nanotechnologies, are currently developing these devices (see Chapter 6, Nanobiomedicine).

Aligned nanotubes have also been used as cathodes in demonstration models of flat-panel displays and television sets. GE Global Research and a small start-up called Molecular Nanosystems are collaborating to develop nanotube field emission devices, including flat-panel displays and medical-imaging products. Samsung has developed a prototype television set that uses a large array of aligned nanotubes to fire electrons at a phosphorescent screen. Conventional television displays are cathode ray tubes that use a wire to generate electrons in an "electron gun" that sits well back to project to the whole screen, and this accounts for its bulkiness. A nanotube "field effect" display would employ thousands of individual nanotubes that could be aligned in a layer that would appear essentially flat immediately behind the screen. Whether these nanotube devices can be produced at price which is competitive with plasma or liquid crystal displays remains to be seen. Newer, thin-layer polymer LED displays, like those invented by Cambridge Display Technologies (see Chapter 7, Nanoelectronics) may well make all of these monitors obsolescent.

Carbon nanotubes have also been used to make "supercapacitors". An ordinary capacitor is an electronic component that stores up charge and then releases it all at once. A supercapacitor is the same thing, only more so. Ning Pan, a professor at UC Davis, his postdoctoral researcher Chunsheng Du, and Jeff Yeh of Mytitek Inc., a small company in Davis, CA, have prepared created supercapacitors from carbon nanotubes aligned on nickel foil.

Conventional, or Faraday capacitors, store electrical charges between a series of interleaved conducting plates. Because every carbon atom in a nanotube is on its surface, nanotubes provide a huge surface area on which to store and release energy.

The devices from Davis produced a power density of 30 kilowatts per kilogram (kW kg^{-1}), compared with 4 kW kg^{-1} for the most advanced devices currently available commercially. Supercapacitors may find application in hybrid electrical or fuel cell powered cars that need a large burst of energy to start their movement.

Carbon nanotubes are fluorescent, and this property makes them suitable a molecular sensors. A feasibility study by Michael Strano, a professor of chemical and biomolecular engineering at Illinois, has shown that nanotube sensors can be used to detect glucose. Glucose sensors, of course, are of paramount importance to diabetics.

"We have developed molecular sheaths around the nanotube that respond to a particular chemical and modulate the nanotube's optical properties," says Strano, "...and, because nanotubes won't degrade like organic molecules that fluoresce, these nanoparticle optical sensors would be suitable for long-term monitoring applications."

Carbon nanotubes have also been used as components of artificial muscles by Ray Baughman, of the University of Texas, who is now working on fuel cell-powered synthetic muscles that may eventually come to the aid of now-paralyzed humans.

CalTech researchers have used nanotubes as components in a nanoscale valve, that may eventually be used in implanted drug delivery devices. Another potential use of the device is for very high-resolution inkjet printers.

Researchers at Rensselaer Polytechnic Institute (NY) and Banaras Hindu University (India) have produced carbon nanotube filters that efficiently remove micro- and nanoscale contaminants from water. The cylindrical filters are composed entirely of radially aligned nanotubes, manufactured in a novel manner using "spray pyrolysis". To fabricate them, the researchers spray a ferrocene/benzene solution into a tube-shaped quartz mold and then heat the mold to 900 °C. The nanotubes grow outwards on the walls of the removable tubular template, using iron particles derived from the ferrocene as a catalyst. The resulting nanotube filters are very strong, heat-resistant and reusable. The researchers demonstrated that they could filter out poliovirus (25 nm in diameter) from water. Another potential use of the filters is the removal of heavy hydrocarbons from petroleum products, for example, in the production of high-octane gasoline.

Scientists at CSIRO Textile and Fibre Technology in Australia have figured out how to spin carbon nanotubes into yarn. The extreme strength and high heat conductivity make this fabric ideal for applications such as body armor. Eventually, it is thought, the nanotube fabric might be able to incorporate environmental sensors.

Potentially economically important uses of nanotubes are as additives in polyurethane to make a very wear-resistant surface coating or to make carbon fiber composites used in very strong, lightweight bicycle parts, with potential applications in aerospace. Prosaic though they may be, some high-tech engineering is required to make nanotubes stick within the matrix materials. Zyvex, with its Nanosolve™ product line, is pioneering these areas. Other companies, including QuinetiQ, Foster-Miller, Inc., and Carbon Nanotechnologies, Inc., have been awarded grants to develop conductive sealants and caulks for military aircraft that incorporate carbon nanotubes.

One of the more unlikely uses proposed for nanotube fibers is to create a ribbon some 100000 km long, to be hung in orbit, and used as part of a "space elevator." Such an elevator was proposed more than 50 years ago in a Russian journal and has made its way into science-fiction, first in a novel by Arthur C. Clarke in 1978. Since then, the space elevator has become a staple in science-fiction, as obvious and necessary to aficionados as wormholes and warp drives. However, not until the advent of nanotubes was there a material that had sufficient strength to make the space elevator possible in reality. The Space Elevator is discussed in more detail in Chapter 8, on Grand Challenges.

Not all nanotubes are made out of carbon. Nanotubes have been created using carbohydrate molecules and other organic chemicals, and natural nanotubes have been found in a certain type of clay. The physical properties of such nanotubes depends upon their constituent molecules.

Dendrimers

Dendrimers are a class of molecules invented, named and patented by Donald Tomalia while he was working at Dow Chemical in the 1970s – practically the Dark Ages as far as nanotechnology is concerned. Today, over 1000 U.S. patents reference the word dendrimers, the name of which is derived from dendron, the Greek word for tree. Perhaps not coincidentally, Tomalia is a tree farmer in his spare time.

For three days in May in 2005, scientists from eighty-two countries made a pilgrimage to the Mecca of dendrimers, otherwise known as Mt. Pleasant, Michigan, where Tomalia lives, and where he has established his company, Dendritic Nanotechnologies. The researchers came for the Fourth International Dendrimer Symposium.

Dendrimers are simply branching polymers. Most polymers that we are familiar with, like nylon or plastics, make long, single fibers, starting with simple organic substrate. Many biological polymers, like DNA, are also more or less linear polymers.

The dendrimer substrate – the dendron – has at least one extra branch point, and this allows it to be extended in three dimensions. Complex carbohydrates are an example of a biological form of branching polymers; these have the property of filling lots of three-dimensional space. An example is cartilage, which is composed largely of complex carbohydrates.

Dendrimers of the kind that Tomalia first created start with a seed molecule, for example, ammonia. Basic precursor units (e.g., diamine or acrylic acid) bind to the ammonia molecule to form the first-generation shell of a dendrimer. The ends of each of the monomers become binding sites for more monomers, creating a second-generation shell, resulting in a structure that branches out in three dimensions like a spherical tree (Fig. 29). More and more layers are added with the generation of polymers. A dendrimer, maintains Tomalia, looks like an onion at the molecular level, and grows (is synthesized) like an onion from the inside out, with each new layer adding a nanometer or so to its diameter. As it becomes three-dimensional, it takes on a globular shape, much like many proteins do. In fact, dendrimers are usually about the same size as proteins, although, in theory, they can be grown much larger. Moreover, like proteins, dendrimers can be created as module that self-assemble into higher-order structures.

Dendrimers are one to way to create molecularly defined, three-dimensional objects – one of the cardinal dreams of molecular nanotechnology. Individual dendrimers are, of course, sub-microscopic molecules. However, they can be designed

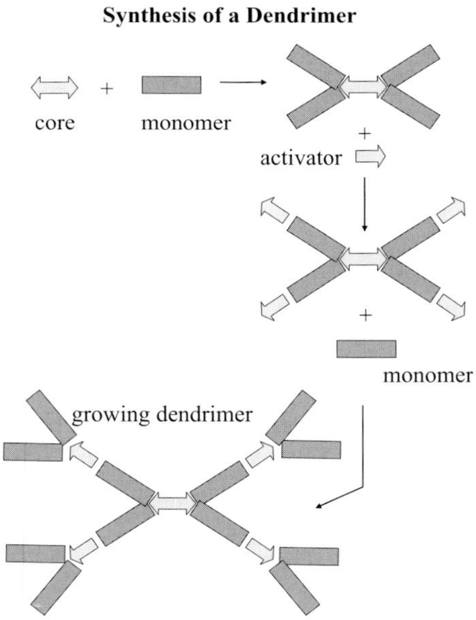

Synthesis of a Dendrimer

core + monomer → activator → monomer → growing dendrimer

Figure 29 Synthesis of a generic dendrimer. A core molecule is reacted with a monomer to produce a four-way branching molecule. This branched molecule in turn is reacted with an activator, which serves as a connector to more monomers. This gives a molecule with eight branch sites. The growing dendrimer would then again be reacted with more activators and monomers, which would double the branch sites each new generation. As the dendrimer grows, it tends to take on a globular shape.

so that they spontaneously form films or develop into three-dimensional shapes like rods. Macroscopic aggregates of dendrimers have been demonstrated. So a chemist, building molecules the old-fashioned way, can nevertheless create something large enough to be visible that is defined down to the individual atom, just like old Mother Nature herself.

The visitors to the fourth annual symposium heard about thiophene dendrimers, polyurethane dendrimers, acrylamide dendrimers, starburst dendrimers, dendrimers decorated with crown ethers, dendrimers coated with fullerenes, dendrimers anchored by buckyballs, dendrimer gels, dendrimer coatings, etc. We won't go in to the actual chemical structures of these curious polymers here. A more important question is: What can you do with a dendrimer? Table 8 provides some answers, which are discussed in the following sections.

Table 8 What can you do with dendrimers?

Company or Research Group	Application
V. Percec, University of Pennsylvania	Liquid crystal lattice
Suslick and Zimmerman, University of Illinois, Champagne-Urbana	Artificial antibodies (aptamers)
G. Thoma, Novartis Pharma	Antibody-blocking agent
J. Baker, University of Michigan	Biosensor, drug delivery
Starpharma	Antiviral and antimicrobial drugs
Cambridge Displays	Phosphorescent dendrimers for LEDS

Virgil Percec, at the University of Pennsylvania with collaborators at University of Sheffield, created a liquid crystal lattice out of dendrimers. The lattice is one of the most complex ever made via self-assembly, where molecules organize themselves into larger structures. This structure contained hundreds of thousands of atoms.

To create these large nanostructures, Percec and his colleagues started with a carefully designed, well-defined and highly branched dendron. When thousands of these molecules come together, they organize themselves, unaided, into discrete microscopic spheres.

In the liquid crystal phase, each sphere consists of 12 tapered dendrons linked at their narrow end. Percec and his colleagues observed 30 of these globular structures arrange themselves into a tetragonal lattice, the repeat unit of which is a rectangular prism containing 255 240 atoms and measuring 169 × 169 × 88 Å. The size of the repeat unit is similar to the crystal form of some spherical plant virus particles. Percec's group is now tweaking the structure of their dendron molecules in hopes that they might evolve into hollow spheres.

Biological researchers have big plans for dendrimers, ranging from uses as drugs, for drug delivery, to uses as artificial antibodies. One company, Qiagen, already markets a type of dendrimer for research investigations that promotes the delivery of DNA into cells. Dendrimers coupled to gadolinium have been used as contrast agents for use in MRI imaging. According to Tomalia, these have been used for over a decade in animals, with no apparent side effects.

Antibodies and Anti-Antibodies

A type of artificial antibody (aptamer) has been created by a team of chemists at the University of Illinois at Urbana-Champaign led by Steven C. Zimmerman and Kenneth S. Suslick, using dendrimers.

"In essence", said Suslick, "... we are molding this dendrimer around our template and creating a rigid cast that functions like a molecular lock for a molecular key." He likens it to the "lost wax process used in metal casting." Molecular recognition by dendrimers could be important for creating catalysts, sensors, and medical diagnostics.

Gebhard Thoma of Novartis Pharma AG in Basel have carried out the opposite sort of experiment, using dendrimers to block the effect of natural antibodies. αGal is a carbohydrate that is present in the cell walls of all mammals except humans and Old World monkeys. Pre-existing antibodies raise an immune response when they bind to αGal, signaling the presence of foreign cells in the body. This reaction is responsible for the hyperacute rejection that occurs when a foreign organ is introduced into a human body. The αGal-anti-αGal prevents pig organs from being transplanted into humans. If pigs could be used, it would be a godsend, given the shortage of human organs.

The dominant anti-αGal antibody is a polyvalent IgM molecule, which means it has a number of binding sites for αGal molecules at once. This makes it difficult to inhibit. Thoma and colleagues suggested that dendrimers with αGal groups in their outer layer might be particularly effective at inhibiting the antibodies, because they would have many different αGal sites for the IgM to bind.

The researchers made several generations of dendrimers from a branched aromatic building block, to each of which they attached αGal units at the tips of the branches. Neither free αGal molecules nor the first-generation dendrimers effectively inhibited the IgMs, and the second- and third-generation molecules are both highly potent as IgM inhibitors. The researchers tested the most effective of their blocking agents in cynomolgus monkeys. Within 5 minutes of injecting the nanoparticles, the IgM antibodies were reduced to 20 % of their original level.

Antibodies can also be attached to the outside surface of dendrimers to target these to specific cells, for example tumor cells. Dendrimers, in this case, can serve as specific carriers of chemotherapeutic drugs, such as cisplatin. Targeting these drugs to tumors helps to prevent their toxic effects on normal cells.

Aiding the delivery of drugs is the development of "exploding dendrimers." Because a dendrimer is essentially an oversized molecule with identical subunits, it has been possible to develop dendrimers that catalyze their own destruction.

Such dendrimers could release their content of drugs all at once, in response to a triggering event.

Dendrimers as Sensors

The spherical surface of a dendrimer can be used as a substrate to attach a variety of functional molecules. In studies to be conducted with the NASA grant at Michigan's Center for Biologic Nanotechnology, under James R. Baker, fluorescent molecules will be attached to the outside of dendrimers. Simply ingesting or inhaling a solution of these tailored molecules will allow them to find their way into white blood cells in the bloodstream. The particular fluorescent molecules being used glow only in the presence of proteins associated with cell death. Once inside the cells, the dendrimers would become real-time monitors of radiation exposure or infection, both of which result in the death of white blood cells.

The idea is to develop a laser-based retinal scanner that can detect the fluorescent dendrimers inside the white blood cells as they pass through blood vessels in the retina. Because blood capillaries are very small, white blood cells must pass through them in single file. Therefore, each blood cell can be separately scanned for fluorescence as it passes. Another parameter that may be examined using dendrimeric nanoparticles is the concentration of calcium within the cells, a measure of cell health. Damage by radiation, for instance, causes calcium efflux.

"We can get a platform that we can target that's less than 5 nanometers in diameter. It provides a very nice scaffold and one that certainly can get through vascular pores and into tissue more efficiently than larger carriers," says Baker.

Dendrimers as Drugs

Starpharma is an Australian firm that has received approval from the U.S. Food and Drug Administration to begin clinical trials of a dendrimer-based drug called Vivagel as an HIV preventative. The company believes that the drug may also have value as a therapeutic, but it is pursuing the preventative indication first, because it is a lower regulatory bar to clear. Starpharma has patent applications on the use of dendrimers as antibacterial and antiviral agents. According to Starpharma's development manager, Tom McCarthy, dendrimers act like "molecular Velcro"; they attach themselves to the surface of pathogens at multiple sites and prevent the bugs from interacting with their cellular targets (Fig. 30). Such tailored dendrimers can interact with viruses at multiple sites in a manner that mimics proteins and other natural molecules

VivaGel also succeeded in preventing infection in monkey trials using a humanized strain of simian immunodeficiency virus. Studies conducted at the University of Washington found that a single vaginal application of the gel protected 100 % of the macaques exposed to the virus. The gel not only prevents adhesion to healthy cells, but also incapacitates the virus even if it does enter the cell. Human clinical testing of the gel has recently begun.

Starpharma is also pursuing the possibility of dendrimers drugs versus a large number of pathogens, including bacteria involved in sexually transmitted diseases as well as respiratory and tropical viruses.

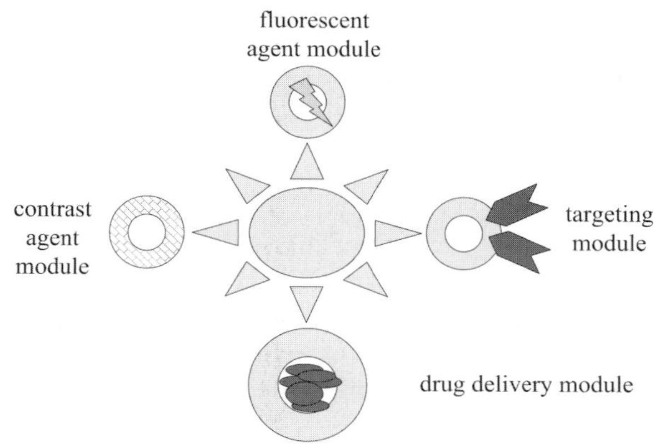

Figure 30 Dendrimer smart drug. Each of the four dendrimers surrounding the core dendrimer has a different function. The fluorescent agent allows the presence of the drug to be detected in the body by fluorescence. The contrast agent allows the drug to be detected by MRI. The targeting module has antibodies or other tissue-specific recognition factors that direct the drug to a specific location in the body. Finally, the drug delivery module is loaded with a therapeutic drug.

The National Cancer Institute (NCI) would like to treat cancer in ways that are much smarter than current chemotherapeutic agents. These new agents would be multifunctional. For example, one module would target the device particularly to tumor cells, whilst another module would contain a contrast agent that would make it visible by MRI imaging, X-rays, or cause it to glow with fluorescent light in response to laser stimulation.

Another module would contain a cell-killing drug that could be released by light activation, or possibly by magnetic fields. This is not just a dream, it is already a field of investigation. In particular, NCI is funding James Baker of the University of Michigan to research these possibilities. In Baker's view, these modules would be composed largely of dendrimers.

Dendrimeric Displays

A small British company called Cambridge Display Technology (CDT) is incorporating dendrimer technology into its polymer light-emitting diodes (PLED)s. These polymers emit light of various bright colors in response to electrical stimulation. By combining the benefits of polymer and dendrimer systems, CDT has found that its systems can be further improved. For example, by preparing polymer–dendrimer hybrid materials, its diodes can be more efficient and operated at lower voltages. CDT believes that dendrimers PLEDS have a particular advantage for battery-operated systems. Dow Chemical also is working on polymeric light-emitting diodes using dendrimers, some of which are produced under license by Dendritech, which is probably the largest commercial manufacturer of dendrimers.

Dendrimers are also being used, according to Michigan-based Dendritech, as low-level additives to increase the adhesion of inks and toners to glass, metal, and plastics. The company expects new applications to be found for dendrimers in nano-lithography, electronics, photonics and chemical catalysis. Solar panels and batteries are among the products that could use dendrimers in the future.

Quantum Dots

Quantum dots are crystals that are so small that their properties are subject to quantum effects. They are sometimes called zero-dimensional objects, although this is not strictly true. Still, they are far too small to be seen without an electron or atomic force microscope.

Given the limited resolution of our natural vision, it is sometimes difficult to appreciate the true granularity of the world. Below is a quote from the great physicist Erwin Schrödinger [2] which illustrates this point:

> Why are atoms so small?... Every little piece of matter handled in everyday life contains an enormous number of them. Many examples have been devised to bring this fact home to an audience, none of them more impressive than the one used by Lord Kelvin: Suppose that you could mark the molecules in a glass of water; then pour the contents of the glass into the ocean and stir the latter so as to distribute the marked molecules uniformly throughout the seven seas; if then you took a glass of water anywhere out of the ocean, you find in it about a hundred of your marked molecules.

A quantum dot is not as small in diameter as an atom, but it doesn't have to be much larger. The smallest created so far contains only three atoms, the same as a molecule of water. And some quantum dots are a sort of virtual matter composed entirely of confined electrons which, with their insignificant mass, manage to cohabit in a crystal formed of ordinary atoms.

A quantum dot is a nanoparticle with decidedly weird properties. Even the name sounds exotic. So what is it? To explain, we will have to do spend a little time in the chemistry class. As chemistry is tedious enough to most people, I will avoid, as much as possible, adding physics to the curriculum.

Chemical elements are made up of atoms, and these atoms are composed of neutrons, protons, and electrons. Neutrons and protons are in the center, the nucleus, of the atom, whiles electrons orbit around the nucleus, like planets around the Sun. Neutrons add mass to the atom, but the properties of any given element are mainly the result of the electrostatic attraction between negatively charged electrons and the positively charged protons. Different elements have different numbers of electrons and protons; hydrogen, the lightest element, has one of each; uranium, the heaviest natural element has 92.

The main rule governing the elements is that the number of electrons is equal to the number of protons. Like any rule, this one is made to be broken. The element hydrogen, for example, is usually found in water as a loose proton, minus its electron. Adding or subtracting electrons to any atom leaves a net negative or positive charge on the atom, making it an ion.

Now all electrons are created equal but, as with real estate, location, location, location is important. The electrons closest to the nucleus are more tightly bound; just as the gravitational pull of Mercury feels the pull of Sun's gravity more greatly than does Pluto, in the far reaches of the solar system.

There are quantum mechanical rules governing the orbits of electrons around the nucleus, which we will not go into in detail about here. Suffice it to say that each orbit (or shell) has a certain number of slots or holes into which electrons can be put. As you can imagine, there is more space available for electrons in the outer shell than there is close to the center.

In an element that has most of its outer shell empty, the electrons are free to move around a lot. Atoms of these elements can also borrow electrons from another atoms, since they have lots of space available. Such elements – like copper, silver and gold – are metallic and are also called conductors, because they permit the free flow of electrons from atom to atom, throughout a crystal or down a wire. In other words, they support electricity.

Elements such as sulfur, which have all or nearly all of their outer shell filled, are non-conductors or insulators. Elements such as silicon, in which the outer shells are about half-full, have behavior that is somewhere between conductors and insulators and are therefore called semiconductors.

Back in the 1980s, researchers began to wonder what would happen if one were to trap electrons in an island of conducting or semiconducting material that was surrounded entirely by an insulator. They made traps out of very thin layers, a few nanometers thick – something the computer chip fabrication industry was just learning to do. The result was that electrons could move freely in two dimensions, but in the third dimension they were trapped by something called "quantum confinement"; the distance they could travel in the third dimension was smaller than the wavelength of the electron, so they were effectively trapped in a two-dimensional plane. These traps were called quantum wells. These wells have interesting properties; application of a voltage across them causes the release of light waves of very precise wavelengths. These quantum wells are the basis for the curiously intense, tiny lasers that people now attach to their key chains.

Scientists were not satisfied with two-dimensional confinement; they eliminated another dimension by trapping electrons in "quantum wires." Electrons could travel in one dimension in elongated conducting or semiconducting pathways, but the other two-dimensions were cut off by insulators. These became the basis of very powerful lasers that could cycled (be turned off and on) as much as 40 gigahertz (cycles per second), and they have application in communications technology.

Finally, researchers' curiosity drove them to create the ultimate in cruel confinement – the quantum dot. They created an island of conducting material so small that the electrons could not travel in any direction. Electrons injected into these dots had nowhere to go, but they didn't precisely sit there, either. In fact, they smeared out into standing waves of probability, very much like the orbitals of an atom. But these electrons did not orbit a particular nucleus as all the protons within the island were already matched with electrons of their own. So, these electron orbitals created within the quantum dot a sort of "artificial atom" of their own. These orbitals are determined by the number of electrons and the space into which they are confined. Artificial elements created by quantum confinement have variable "chemical" properties just like real ones. These tunable properties led Wil McCarthy [3] to promote the idea of programmable matter that can be changed at will from one form to another by adjusting voltages – artificial lead could be transmuted into artificial gold, the ultimate in alchemy. How practical this would be is another matter. Real gold is probably cheaper, in the near term, but nevertheless McCarthy has applied for a patent on the idea.

Colloidal Quantum Dots

Natural quantum dots also exist in the form of small clusters of metal atoms, for instance colloidal gold. If the cluster is small enough, then the electrons are subject to quantum confinement defined by the shape and size of the cluster. In this case, the conduction electrons are shared by the whole cluster as if it were a single atom. These electrons have unique and defined orbits within the quantum dot.

When such small crystal dots absorb light, they do so in a very precise manner. Electrons that absorb the light energy are promoted to a vacancy in a higher energy orbit. When the electron decays to its original orbit, it emits a photon of light of a particular wavelength corresponding to that loss of energy – a type of fluorescence.

In a conductor or semiconductor, excitation of an electron by a photon of light in one of the inner shell electrons (called valence electrons) will promote it to a vacant slot in the outer shell or conductance layer. This leaves a hole in the valence band, which is normally always full. The hole/electron pair is called an *exciton*. The physical separation between the distance between the electron and the hole is called the *exciton Bohr's radius* (after Niels Bohr, a quantum mechanics pioneer who first calculated the radius of a hydrogen atom).

In a colloidal quantum dot, the size of the particle is smaller than the normal Bohr's radius of the compound, setting up a situation called *quantum confinement*. In this case, the jump in energy level of the electron is limited by the size of the particle. A consequence of quantum confinement is that the wavelength of light – and therefore its color – depends on the size of the particle.

For instance, Robert Dickson of Georgia Tech and his colleagues have created gold quantum dots which are made up of precisely 5, 8, 13, 23, or 31 atoms. Each size fluoresces at a different wavelength to produce ultraviolet, blue, green, red, and infrared light, respectively. The fluorescence energy varies according to the radius of the quantum dot, with the smallest dots being the most efficient.

This is intuitively kind of weird. Imagine a red brick. Now imagine a brick made up of the same material that is twice as big. Should the second brick be blue instead of red? But most bulk objects are not fluorescent; the red from the brick is just the average color of the light we see reflected back at us. Quantum dots, on the other hand, are not really colored, at all. As they are smaller than the wavelength of visible light, they are actually invisible. But they blink on and off like Christmas lights, emitting fluorescent light that can be seen.

The researchers create their dots by reducing gold salt in the presence of polymers that create dendrimers. By controlling the relative concentrations of gold and polymer and the rate of dendrimers generation, they are able to quantum dots encapsulated in dendrimers with very defined sizes.

Another professor, Munir Nayfeh, a University of Illinois, has fabricated nanometer-scale quantum dots containing only 29 silicon atoms, which glows a bright fluorescent blue. A 1.67-nm particle with just 123 atoms of silicon emits a bright green under UV excitation. Slightly larger particles produce red or yellow light. The UV excitation raises electrons to higher energy levels that they can sustain in their confined state, so they emit that extra energy as light in the visible range.

Early investigations on colloidal or crystalline quantum dots were actually aimed at developing better solar panels. Two research groups discovered that nanoparticles of semiconductor materials have a unique optical property that became known as quantum confinement. Alexander Efros and Alexie Ekimov were at the Yoffe Institute in Leningrad, Russia, and Louis Brus and his team were at Bell Laboratories. Both teams discovered that tiny crystals of cadmium selenide would fluoresce different colors when hit with light, depending upon the crystal size. Two of the Bell Labs investigators, Moungi Bawendi and Paul Alivisatos, went on to Massachusetts Institute of Technology in Cambridge and the University of California in Berkeley, respectively, where they made the nanoparticles water-soluble and devised zinc-based inorganic shells to enhance their fluorescence. Both men are on the scientific board of Quantum Dot Corp., based in Hayward, CA, which licenses the intellectual property from their discoveries.

Quantum dots are now the subject of over 700 issued U.S. patents since 1986, and over 800 pending patents since 2001, according to John Oliver, an industry analyst for BCC, Inc. Some industrial behemoths are interested in quantum dots, including Micron Technology, IBM, Texas Instruments, and Motorola. However, small entrepreneurial companies hold some of the more innovative patents – companies such as Quantum Dot Corp., BioCrystal, Inc., or Nanosphere – not exactly household names – yet.

Quantum Dot was founded in November 1998 by Joel Martin and Bala S. Manian. Martin is the co-founder and former CEO of a company called Argonaut Technologies that makes instruments for the discovery of drugs and new materials. Bala Manian, a native of India, made his name in the U.S. developing techniques that allowed filmmakers such as George Lucas to insert special effects into movies like *Indiana Jones* and *The Return of the Jedi* using computerized digital imaging. Quantum Dot is his seventh start-up company.

In contrast to the gold dots made by Dickson, current commercially made quantum dots are much larger, containing hundreds or even thousands of atoms. These are usually made from semiconductors such as cadmium selenide or lead selenide. These materials are doubly disadvantageous in that they are highly toxic and insoluble in water. The manufacturers counter this problem by coating the dot material in zinc sulfate to prevent the cadmium selenide crystal from dissolv-

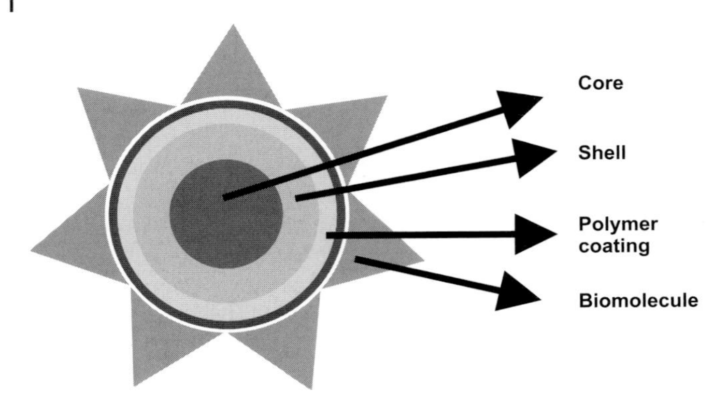

Figure 31 Commercial quantum dot architecture (courtesy of Quantum Dot Corp.) The core quantum dot material is made from cadmium selenide. This is coated with a zinc sulfate shell. A polymer coating makes the dot soluble in water and may be used to attach biomolecules such as antibodies or oligonucleotides. The color of the quantum dot depends on the size of the core, usually from 2 to 10 nm.

ing, and then coating that with a water-soluble (hydrophilic) polymer. The final coating has the advantage that functional biomolecules can be linked to it (Fig. 31).

Quantum Dots in Biology and Medicine

As yet, the major use for quantum dots has been to replace the florescent labels used in biological experiments. For years, cell biologists have attached fluorescent molecules (fluorophores) such as fluorescein (which glows green) or rhodamine (which glows red) to mark certain molecules within a cell. Typically, in these experiments, the fluorophore would be attached to a biological molecule, like an antibody, that has recognition properties. An antibody made to a particular cellular protein – for example, the protein tubulin – attached to a fluorophore allows you determine where that protein is located within the cell. Tubulin is the major protein in the microtubules, elongated structures that make up part of the "skeleton" of the cell; a fluorescently-labeled antibody to tubulin allows you to see these otherwise invisible structures lit up like Christmas trees within the cells. A drawback of organic fluorophores is that they "bleach" within a minute or two; the Christmas lights fades, so you have to take a picture very quickly to document your results.

Quantum dots do not bleach and they are also about 10- to 20-times as bright as commonly used organic fluorophores. This provides biologists with a whole new range of experiments that they can perform.

Quantum Dot Corp has released its first products, called Qdots, which are cadmium selenide quantum dots that fluoresce red, yellow, or green. Qdots may also be modified by putting biological molecules on the surface for use in various biological experiments and assays. Evident Technologies, of Troy, New York, makes similar products.

John Frangioni of Harvard Medical School and colleagues used quantum dots to locate "sentinel" lymph nodes that frequently capture metastatic cancer cells; sentinel lymph nodes are often removed for cancer diagnostic screening, but they can be difficult to locate. The brightly lit dots can easily seen through the skin of living mice. Similar procedures may be useful in diagnosing breast cancer in humans.

"The size of the quantum dots turns out to be ideal for getting into the lymph system and then getting trapped in the sentinel lymph node," says Andy Watson, vice president of business development at Quantum Dot.

A limitation of the fluorescent dots is that visible light frequencies would only be useful near the surface of the skin. A newly developed quantum dot nanocrystal, that emits near-infrared light has been synthesized in the MIT laboratory of Moungi Bawendi, who is a scientific co-founder and advisory board member of Quantum Dot Corp. (QDC). Near-infrared penetrates tissue much more efficiently than visible light. Although it cannot be seen by humans, near-infrared light can be detected with instruments.

"Sentinel lymph node mapping has already revolutionized cancer surgery. Near-infrared quantum dots have the potential to improve this important technique even further," says Frangioni. His laboratory has also developed methods to target quantum dots to specific cancer cells, possibly allowing the early identification of tumors or the localization of metastases.

Researchers at Rockefeller Institute have used quantum dots to study living cells in whole organisms, such as mice, frog embryos, and slime mold. Because the fluorescence properties of the quantum dots is long-lasting and not easily degradable in the body, it is possible to study the movement of cells, or even the movement of proteins within the cells, over periods of several days. As many as a billion quantum dots have been injected into a frog cell, without evidence of toxicity.

Quantum Dots in Electronics

Eventually, it is thought that quantum dots will find many applications in optoelectronics. Prototype quantum dot containing light-emitting diodes and optical transistors have been created, as well as lasers that can cycle very quickly for use in optical communications and data transmission. The original goal of colloidal quantum dot research was to adapt these devices for use in solar panels, and as costs come down this may emerge as a commercially viable application.

Quantum dots have been proposed as a way of increasing the efficiency of fluorescent lighting, which is already much more efficient than incandescent lights. Conventional fluorescent lights are filled with a gas that emits ultraviolet (UV) light when current is passed through the tubes. When the UV light hits phosphors coating the tube, the phosphors emit fluorescent light at visible wavelengths. Quantum dots can be tuned to the same thing. In a prototype quantum light bulb developed at Sandia National Laboratory in Los Alamos, a quantum well supplies the UV light when it is pumped with energy delivered by a laser. The UV light, in turn, stimulates quantum dots to do their thing. Although the dots emit monochromatic (single color) light, by mixing colors, white light can be approximated. The quantum light bulb promises to be much more efficient at generating light from electricity than fluorescent bulbs.

Quantum dots are already being used in experimental "quantum computers" that some researchers expect to transcend conventional computers, though the conventional wisdom is that commercial-grade devices are at least twenty-five years in the future. Quantum computing will be described more fully in Chapter 10.

References

1 Iijima, S. Helical microtubules of graphitic carbon. *Nature* 354: 56 (**1991**).

2 Schrödinger, E. *What is Life: The Physical Aspect of the Living Cell.* Cambridge: at the University Press, New York: The MacMillan Co., **1947**.

3 McCarthy, W. *Hacking Matter.* Basic Books, New York, NY, **2003**.

Chapter 6
Learning from Old Mother Nature

In the 21st century, men have learned finally to make simple machines at the nanoscale. Old Ma Nature, on the other hand, turns out very complex nanoscale machines, like ribosomes and mitochondria, by the gazillions daily and has been doing it for billions of years. How many billions of years? Well, that is still an object of intense debate. In the 1950s and 1960s, it was accepted as an article of scientific dogma that life had formed spontaneously on Earth about 2 billion years ago. Since then, fossils of algae and bacteria have been found that suggest that life appeared on Earth shortly after the planet cooled, about 3.5 billion years ago. For many scientists, the rapid appearance of life on Earth seems to be too quick to have been spontaneous. Nobel laureate Francis Crick, who along with Joseph Watson discovered the structure of DNA, has argued that Life may, in fact, have been planted on Earth from somewhere else in space in a process that he calls "Directed Panspermia [1]." The British astronomer Fred Hoyle remarked that "... the formation of life-forms, with all their complexities, from the random couplings of prebiotic chemicals is about as likely as a Boeing 747 being assembled by a tornado flying over a junkyard." Physicist and Biblical scholar Gerald Schroeder has advanced the novel theory that the complexity of early life actually argues for an Intelligent Creator [2]. We introduce this debate not to wander off into metaphysical tangents, but only to make a point of our own: that life relies on exquisitely tuned tiny machines, and that Nature has had a long time to perfect them – 3.5 billion years at a minimum, and possibly much more. That being the inarguable case, what can we learn (or borrow) from Mother Nature?

Biomimetics or biomimicry is what we call it when men use Nature as a model for items of manufacture. The case most often used as an example is the ubiquitous fastener Velcro, which was consciously patterned after the grappling hooks developed by cockleburrs to grab hold of passing animals or people's clothes as a means to disseminate themselves more widely. Many wonderful devices, many of them too small for easy examination, are out there for the clever researcher to reverse-engineer from Nature, keeping the patent rights for himself. Ma Nature graciously provides her intellectual property free-of-charge.

Some biomachines are relatively simple. An example is the microtubule, which is a kind of railroad track that cells use to transfer goods from the nucleus to the

The Nanotech Pioneers. Steven A. Edwards

ISBN: 3-527-31290-0

cell membrane or to various organelles (the organelles are the guts of the cell, analogous to the organs of the body). The microtubule is comprised almost entirely of a single protein called tubulin. In solution, tubulin molecules self-assemble, snapping together like so many Lego blocks, without the aid of tiny fingers.

In the pages that follow, we will examine a few of biomachines that researchers are already working on to either imitate or to revise: the gecko's foot; the eye of the starfish; the tough shell of the abalone, the myriad silicon forms of diatoms; and the rapidly responsive membranes of nerve cells.

The Gecko's Foot

Geckos are a family of marvelous lizards of vastly varying coloring and habitat, comprising about 1000 species. Unlike most lizards that are active during the day, Geckos are mostly nocturnal. Their most amazing trait is their ability to stick to almost anything. Geckos can climb smooth Plexiglas surfaces and can even walk on the ceiling, hanging out as they please, apparently immune to the law of gravity. High-speed video of geckos in the process of running show that their motion is practically indistinguishable whether they are climbing up smooth walls or running across the floor.

How does the gecko do it? Researchers have learned that proximity is an important part of adhesion: if you take out the distance between two smooth surfaces, they have a natural tendency to stick—sheets of glass, stacked together frequently are difficult to prize apart without breaking them. Most apparently flat things don't stick together because at the nanoscale, they are not really flat. A nanoscale ant at a picnic trying to traverse the surface of a paper plate would have ridges and mountain ranges of paper fiber to cross.

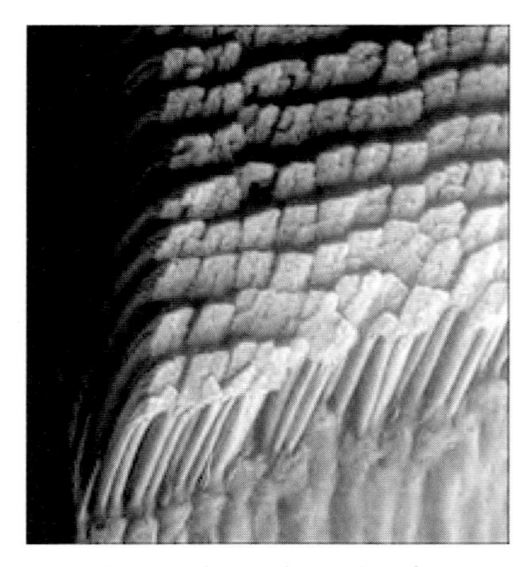

Figure 32 Spatulae on the Gecko's foot imaged by scanning electron microscopy. Reproduced courtesy of Kellar Autumn, Lewis and Clark College, Portland, Oregon.

Each of the gecko's toes are coated with a half million or so bristles called setae; each of these branch further into hundreds of split ends called spatulae. The density of the setae is about three million per square inch. The spatulae are only 200 nanometers wide, small enough to be invisible except with the use of electron microscopes. The surface area of all the spatulae added together, however, is enormous.

At very short distances, most types of matter attract through what are known as van der Waals forces. Named after a 19th century Dutch physicist, these forces are very weak compared to magnetic forces or electrostatic force—the attraction of a negative charge to a positive charge, for instance. But van der Waals forces are additive; put enough of them together and suddenly you have a very potent attractive force. Kellar Autumn's laboratory, at Lewis and Clark College in Oregon, showed that geckos do exactly that with their bristly toes.

If you sneak up on a gecko and pull him off unexpectedly, he may very well leave a leg or two still attached to the wall. A single gecko foot can suspend a 40 lb weight. Presumably, therefore, a gecko could grow to 160 pounds and still stick to the ceiling, if he could get all of his 6.5 million setae to stick at once.

There are exceptions to the ubiquity of van der Waals forces; in the non-stick surface Teflon, van der Waal's forces are negligible. And sure enough, geckos won't stick to Teflon anymore than scrambled eggs can.

The gecko's foot is an adhesive that can let go on command. An amazing property of gecko's setae is that simply lifting the setae to a 30 degree angle, breaking the van der Waals forces one by one, causes them to release their hold. Epoxy-based glue is just as adhesive, but try to pull your fingers apart after you've glued them together.

Another mystery of the gecko's foot, is why the setae is do not get dirty. Anybody who has ever tried to actually use duct tape on ducts under their house will realize that dust is poisonous to adhesives. Geckos never clean their feet, even though they run through the dirt all the time. Holland's lab showed that isolated setae also are self-cleaning; they don't gum like you might expect. Imagine self-cleaning duct tape. Autumn's lab showed that because only a few setae can adhere to any single dirt particle, the particle doesn't stay stuck very long. The dirt sticks instead to whatever the animal is walking on. The gecko leaves a trail of dirt particles behind him as he climbs the wall.

Van der Waals adhesion is largely independent of chemical composition; it is the geometry that matters. In other words, the size and shape of the tips of gecko foot hair are more important than what is made of. This means, potentially, that gecko toes could be made from almost any material, and researchers are already trying their hand at it. Robert Full and Ronald Fearing at UC Berkeley, together with Kellar Autum, have succeeded in making artificial setae. Full, who uses animals has models for robot design, has helped build Mecho-Gecko, a robot that climbs walls.

Andre Geim, at the University of Manchester, who once shared the honor of an Ignoble prize in physics for the feat of levitating a frog with a magnet, has invented what he calls "gecko tape"—a dry, self-cleaning adhesive, that will stick to almost anything and is re-attachable. It is not just weird inventor types mimicking the gecko; Bell Labs is experimenting with a gecko-foot type adhesive to reversibly cement microchips together.

A gecko-like adhesive has the possibility of becoming "one-sided Velcro" that will stick to anything but can still be easily removed – like the adhesive on a 3M Sticky Note, only strong enough to hang a TV set on the wall. One nanotech company, Nanosys of Palo Alto, CA, is working on a commercial gecko-mimicking product that they call "Nano-fur". Nano-fur is made by attaching nanoscale-fibers by the millions to surfaces. Bob Dubrow, Director of Product Development Management claims to have a product in hand that will allow a 150 pound man crawl up a vertical surface and they have used it to stick boxes on a wall in their offices; it's easier than building cabinets.

The Eye of the Starfish: The Optical Network of the Sponge

Have you ever wondered how a starfish in the sea sees? Perhaps you thought, like I did, that a starfish doesn't need to see. Most starfish, it appears, actually do have a single eye at the end of each arm, so a five-armed starfish would have five eyes. Calling it an eye is probably giving it too much credit; perhaps a light-sensitive spot would be more correct. The starfish can detect light and tends to run (slowly) away when things gets too bright.

Scientists from Lucent Technologies' Bell Labs discovered that a creature called the brittlestar (*Ophiocoma wendtii*), however, is a starfish prodigy when it comes to visual feats (Fig. 33). Thousands of calcite (calcium chloride) crystals that line its surface have a dual function, acting both as nasty, sharp armor and as optical lenses for an all-seeing, 360- degree vision, compound eye. The crystals form regular array of spherical microstructures that can be found in the skeleton of brittlestars. The lenses focus light about 5 μm below the surface of the brittlestar's skin. Nerves running through the skeleton underneath the lenses pick up the signal. Thousands of these calcite crystals form a primitive compound eye that covers much of the organism's body and presumably help the starfish to evade predators. Closely related brittlestars that are indifferent to light do not have these lenses.

"This is an excellent example of something we can learn from nature," said Federico Capasso, physical research vice president at Bell Labs. "These tiny calcite crystals are nearly perfect optical microlenses, much better than any we can manufacture today."

The brittlestar is something of a genius optician; it's calcite microlenses expertly compensate for birefringence and spherical aberration – two common distortions of light that lenses must overcome. Bell Labs hopes to make its own version of brittlestar microlenses for optical communications networks. Single crystal lenses patterned at or below micron scale are important components of optoelectronic circuits in advanced electronic, sensory, and optical devices.

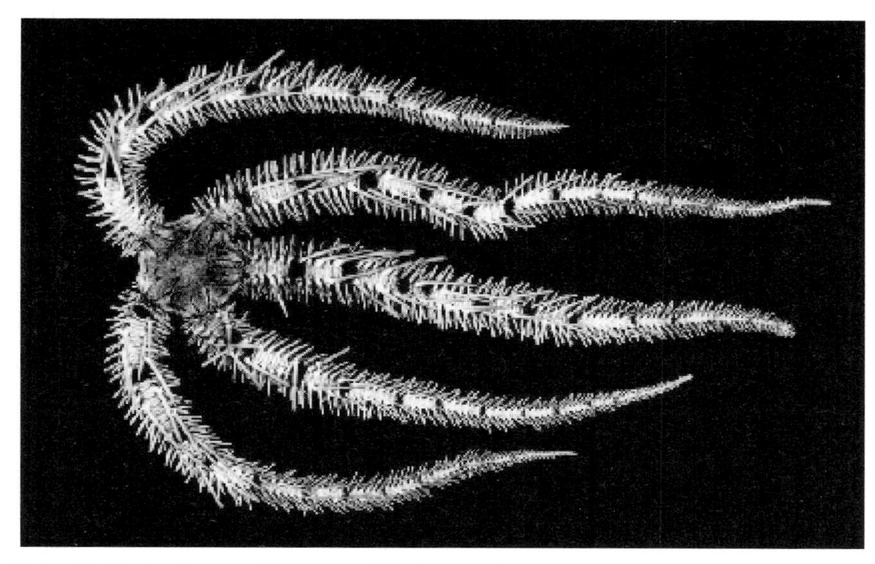

Figure 33 The brittlestar *Ophiocoma wendtii*. Photograph by John E. Miller. The image was provided by Dave Pawson, senior research scientist at the National Museum of Natural History.

Ever since spectacles were invented, high-quality lenses have been created by grinding down a piece of glass. Even the lenses of the Hubble telescope were created in this way. But the lowly brittlestar, like a good nanotechnologist, has developed a whole different bottom-up manufacturing technique, depositing successive layers of calcite onto a patterned organic template to form near-perfect crystalline lenses molecule-by-molecule. Joanna Aizenberg and her team have studied these biomineralization techniques and have been able to fabricate single crystals of calcite with patterns that are approximately one-tenth the diameter of a human hair – not as good as the starfish, but working up to it.

The brittlestar is not the only critter on the sea floor with some optical engineering expertise. Bell Labs has also found that certain types of sponges pioneered the development of optical fibers long before man.

The sponge in question, called poetically the Venus Flower Basket (or less poetically, *Euplectella*), lives deep in the tropical seas and grows to about 15 cm in diameter. It is built around an intricate mesh-like skeleton of glassy silica, and not infrequently plays host to a pair of mating shrimp. At the base of the sponge's skeleton is an inverted crown of fibers, each of which is between 5 and 15 cm long (Fig. 34).

Bell Labs researchers found that each of the sponge's fibers is constructed of distinct layers with different optical properties. Concentric silica cylinders with high organic content surround an inner core of high-purity silica glass; this structure is remarkably similar to industrial optical fiber, in which layers of glass cladding surround a glass core of slightly different composition. The biological fibers of the sponge conduct light beautifully when illuminated.

Figure 34 Venus Flower Basket (*Euplectella*). Image reproduced courtesy of Richard L. Howey, Adjunct Professor of Zoology and Physiology, University of Wyoming.

"These biological fibers bear a striking resemblance to commercial telecommunications fibers, as they use the same material and have similar dimensions," said Joanna Aizenberg, the Bell Labs team leader.

Though these natural bio-optical fibers do not have the transparency necessary for communications network, they are superior in that they are extremely resilient to cracks and breakage. One of the main causes for outages in commercial optical fiber is fracture resulting from crack growth within the fiber. Replacing the fiber is often a costly, labor-intensive proposition as optical cables are frequently underground or draped across the sea floor.

The sponge uses an organic sheath to cover the biological fiber. "These bio-optical fibers are extremely tough," said Aizenberg. "You could tie them in tight knots and, unlike commercial fiber, they would still not crack. Maybe we can learn how to improve on existing commercial fiber from studying these fibers of the Venus Flower Basket."

Another advantage of the sponge's biological fibers is ease of manufacture. Commercial optical fiber is produced with the help of a high-temperature furnace and expensive equipment. Somehow the sponge produces its fibers in a kind of molecular manufacture, essentially chemical deposition at the temperature of seawater.

The Abalone's Shell

Another sea-dwelling nanotechnologist of note is the abalone. A type of mollusk (snail), the abalone has two features that recommend it to humankind. One is that, suitably tenderized and seasoned, the meat of the abalone is very good eating. Second, the "mother-of-pearl" inside surface of shell of the abalone is used in jewelry. These valuable traits have made the abalone increasingly rare in parts of the world where people abound. When I was a child, my cousins and I would go the beach near my grandmother's house in Corona del Mar, CA, and pry these animals off of the rocks with a crowbar. You had to sneak up on them; if they sensed you coming they would squeeze tight against the rock and you couldn't get any purchase with your tool. Once popped off the rock, the muscle was separated from the shell, mashed repeatedly with a tenderizing mallet, then breaded and fried in oil. A little limejuice complimented the flavor deliciously. My father smoked heavily, and every ashtray in our house was an abalone shell. Other shells were shipped eastward for my Chicago cousins to enjoy.

As the population of California increased, the abalone disappeared from the rocks. By the time I had graduated from college, there was scarcely a legal-sized abalone within 30 meters of the surface of the ocean, so thoroughly had they been hunted.

Abalone shell illustrates one of the great advantages of engineering material at the nanoscale – essentially the same material can have widely varying properties according to how it is organized (Fig. 35). The beautiful, iridescent shell of the abalone is made essentially of same material from which crumbly blackboard chalk is made and for that matter, the brittlestar lenses discussed in the previous section. Do you want soft, flaky chalk, a transparent lens, or a hard, beautiful shell? All three are useful and all three are largely made out of calcite. More than 300 different crystal forms of calcite have been identified; these can combine to produce at least a thousand different crystal variations. People have a devil of a time getting the crystal of interest to form, but Nature is very good at it.

Chalk flakes and is easily broken, but abalone shells are more like armor plating. So what is the difference? Calcium carbonate is organized by the abalone along a matrix of protein and carbohydrate, much as the calcium in our teeth is layered into enamel. In particular, the abalone protein controls the growth, shape and eventual size of the calcite crystals. Abalone shell is about 95 % calcite, with the remaining 5 % composed of protein and carbohydrates.

One aspect of abalone shell is its resistance to cracking. A blow to the surface of the shell causes tiny cracks that do not propagate easily; the structure of the shell forces the cracks to go around microscopic tile-like structures – nanoscale masonry, dissipating the force of the blow, as shown in Figure 35(b). This trick makes the shell about 3000 times harder and more resistant to breakage than a single calcite crystal the size of an abalone shell would be.

The elastic glue that holds the abalone shell together fills the crack and immediately begins recruitment of more calcium carbonate to repair the shell. Self-repair is another one of Nature's unpatented properties that engineers would dearly love to engineer into man-made materials.

Mark Meyers at the UC San Diego's Jacobs School of Engineering believes that the abalone shell architecture may be a guide for the development of new bullet-stopping armor. He is also interested in another marvel of evolution, the bill of the toucan. South American bird's large nut- and berry-cracking bill is at once both extremely strong and extremely light-weight.

Meyers and his graduate student Albert Lin have demonstrated that the abalone's shell is the toughest arrangement of its materials that is theoretically possible – a tribute to the engineering prowess of evolution.

Nevertheless, nanotechnologists hope to go the abalone one better – the sea-going snail has challenged many scientists to see if they can improve on its self-assembling hardened armor. After all, the abalone uses only two of the 300 versions of calcite crystals available to make its tiles. As MIT's Angela Belcher has pointed out, it took the abalone millions of years to evolve its shell. But starting with the natural version as a template, a biotechnologist can design, create and isolate multitudes of new calcite organizing proteins in about three weeks.

Figure 35 Mother-of-pearl, nacre. The inner section of the shell of red abalone (*Haliotis rufescens*), that lives along California and Baja California Coasts, consists of nacre, a nano-composite of calcium carbonate ($CaCO_3$) and proteins. The image in (a), recorded using transmission electron microscopy from a thinned edge-on sample, displays the characteristic brick and mortar composite architecture of the biogenic structure. The segmented-laminated composite material provides one of the toughest and strongest materials (by weight) known to engineers, despite the fact that ingredients of the material are rudimentary, similar to chalk and egg-white, respectively. Energy dissipation mechanisms during controlled fracture include co-operative deformation (inset) and highly-tortuous fracture, evident in the fractured surface recorded by scanning electron microscopy (b) (inset is a polarized light microscopy image of the edge-on nacre near an indentation). Figure courtesy of Mehmet Sarikaya, University of Washington, Seattle, WA, USA.

Diatoms: The Original Silicon Chips

The diatoms are a family of single-celled organisms that float on the surface of the sea, surviving off sunlight and the greenhouse gas, carbon dioxide. They are incredibly important to the environmental health of the planet; they are thought to absorb as much carbon dioxide as all the world's rainforests combined. Although diatoms represent only 1 % of the world's biomass, they account for 50 % of all photosynthesis, more than all land-dwelling plants combined. Diatoms also form the bottom of the food chain in the oceans, nourishing directly or indirectly much of the world's sea life.

Diatoms are not exactly plants and not exactly animals; they combine features of both; biologists believe that diatoms are the result of a fusion between a microbe and a red algae. Diatoms are photosynthetic like plants, but have a certain amount of motility (movement), like animals.

Diatoms are usually a few microns in diameter – smaller than a mammalian cell but larger than a bacterium. Some diatoms, however, can be as large as 100 μm, much larger than most of our cells. The diatom's cell wall, called a frustule, is one of the unique features of a diatom; it is composed of silica (essentially glass) and comes in a bewildering variety of very detailed forms, all of them patterned down to the nanoscale. Silica is the dioxide of silicon; silicon is the most abundant element on the planet by weight, and is part of everything from sea sand to window glass to semiconductor chips. Every species of diatoms (estimates of the number of species range from 100 000 to a million) has its own peculiar-shaped frustule: under the microscope these resemble barrels, hatboxes, pill boxes, stars, tear-drops, pincushions, snowflakes, or miniature alien spacecraft. The lowly diatom is an architect that would put Frank Lloyd Wright or I.M. Pei to shame.

"You show a picture of a diatom to scientists," said David Wright, a professor of chemistry at Vanderbilt University, "and you've immediately got your audience. The structure of these organisms is so amazing, so varied. There's nothing in modern science to match it."

Diatom nanotechnology received its start somewhat accidentally when Richard Gordon of Manitoba University was invited to give a talk on diatoms at an engineering conference in 1988. He wasn't an engineer himself, but he knew that engineers are interested in microfabrication, and who does it any better than the lowly diatom?

Right now, we cannot build silica structures that are anywhere near as perfect and detailed as a diatom can, points out Mark Hildebrand, of Scripps Institute of Oceanography. But what we can do is let the diatom do it for us (Fig. 36). Just as biotechnologists have harnessed the gut bacterium *E. coli* to produce protein pharmaceuticals, we can use diatoms to build for us remarkable nanoparticles built of glass. Ultimately, says Hildebrand, we want to be able to manipulate diatoms genetically, to turn them into the sort of lab animal that *E. coli* has become. In this way, perhaps, we will eventually be able to get the diatom to build frustules to order. Building materials at the nanoscale is time-consuming and expensive for the average engineer, but you can grow billions of diatoms in the lab at the cost of a few dollars. All they require is sunlight, sea water and air. Diatoms usually reproduce asexually through cell division, like bacteria, with a mother cell splitting into two daughter cells.

Figure 36 Various types of diatoms arranged into a decorative pattern. Image reproduced courtesy of Klaus Kemp, Microlife Services, Somerset, UK.

To create their exterior frustules, diatoms extract a dissolved form of silica called silicic acid from seawater, importing it into a particular membrane-bound organelle called a silica deposition vesicle. This is where the silicic acid is shaped into a hard shell.

Researchers are beginning to sequence the genome of diatoms, with one aim being to understand how the architecture of the frustule is controlled. Diatoms, like the abalone, use protein molecules to precisely control the formation, orientation, and morphology of mineral crystals (silica in the case of diatoms, calcite for abalones) and thereby optimize the structure for their use. In principle, if we knew and understood the genome of these creatures, we might be able to engineer them genetically to build a silica nanoparticle to order.

"Diatoms can manipulate silica in ways that nanotechnologists can only dream about. If we understood how they can design and build their patterned frustule as part of their biology, perhaps this could be adapted by humans," said Dan Rokhsar, the head of computational genomics at the Joint Genome Institute, which was responsible for sequencing the genome of a diatom called Thalassiosira.

Researchers have been able to isolate some silica-binding proteins from diatoms that crystallize mold silica and mold it into different shapes. Simply adding the proteins to a solution of silica causes small crystals to form. Different proteins cause different shapes, small spheres, cylinders or plates can be formed in this way.

Ken Sandhage, of Georgia Tech, appreciates what diatoms can do, but he is not satisfied with a silica frustule. He wants nanostructures that are biodegradable or have magnetic or electronic properties. He has performed experiments that one observer compared to the action of Star Trek replicators. By baking diatom frustules at high temperatures in the presence of magnesium gas, he was able to effect a molecule-for-molecule replacement of silica with magnesium oxide. He has since repeated the experiment using instead titanium dioxide, which is used commercially in paints, coating and nanomaterials. He predicts that other molecules can be used as well.

It is expected that specially tailored diatom nanostructures synthesized by diatoms might be used in the near future as nanocapsules to deliver drugs, as diffraction gratings in optical devices, as filters and membranes, as high-surface area catalysts, masks for lithographic patterning, and as controlled-shape structures used in composite materials. Already, researchers at Ohio State are using the nanoporous titanium oxide frustules created by Sandhage in gas sensors. Fancier ideas, such as diatom-synthesized gears and widgets will depend on better genetic control.

Natural Nanotubes

The typical nanotech entrepreneur, if such a being exists, would be an academic professor who thinks he or she has discovered a way to capitalize on their research. Or, alternatively, an academic who has figured out a way for Wall Street to subsidize his or her research. Mike Weiner, CEO of Biophan (see Chapter 8) is not your typical nanotech entrepreneur. First, of all, he is not an academic at all and holds no advanced degrees. Second, he is not buried in some narrow field of specialization, but is interested in absolutely everything – which helps to explain how he could be responsible for some seventeen issued U.S. patents. He travels constantly, from meeting to conference to symposium and talks to everybody. Thus, it was not so unusual to find him a few years ago listening to a presentation by the Atlas Mining Co. Its Dragon Mine in Utah is the only commercial site in North America that produces halloysite clay.

Halloysite clay is used to make porcelain objects, fine china, and a few high-tech ceramic objects. Like other clays, it is a mixture of silicon and aluminum oxides. What caught Mike's attention in the presentation was an electron micrograph of the fine structure of halloysite clay. It was just full of these long, narrow tubes, most of less than 100 nm diameter, but sometimes many microns in length. "Nanotubes," thought Mike, "Natural nanotubes." And then, because he is an entrepreneur, that immediately suggested a name for the company he would build to exploit this discovery, "NaturalNano, the name alone is worth $50 million in market cap," he decided. And so, a few months later, he had set up the company and found a long-time associate, Michael Reidlinger, to run it. "The next revolution in nanotechnology," states the company's website, "is quietly forming underground."

Halloysite nanotubes do not have some of the remarkable properties – extraordinary strength and conductivity – that carbon nanotubes have. On the other hand, carbon nanotubes currently cost about $500 per gram to produce, whereas halloysite clay costs about $500 per ton. About 10% of the bulk clay is nanotubes, that still makes them an extraordinary bargain as nanotubes go. (Atlas Mining refers to its halloysite tubes as "microtubules"; however, there is already a well-known biological organelle by that name.)

Halloysite nanotubes do have a tube-like structure and an extraordinary amount of surface area. Close analysis of cross-sections reveal that they are not closed tubes, but are more like scrolls, or rolled paper. There appears to be a charge difference between the ends and the middle of the tubes: this may account for the structural strength of the clay, as the tubes tend naturally to line up perpendicular to one another.

NaturalNano claims to have come up with over 100 potential commercial application for natural nanotubes. The tiny tubes can be filled with liquids, which tend to absorb strongly to the clay. Thus, they are good for "coatings, antifouling paint, antiscalants, pesticides, pest repellents, household and personal products, flavors and fragrances, pharmaceuticals, and other agents," says the company. The company has already investigated the use of natural nanotubes for slow-release perfume. So, the fragrance that you put on for Friday night's date, instead of evaporating before the good-night kiss, can last through the week-end. Similarly, the nanotubes would be good for slow release of cosmetics and possibly drugs. The nanotubes can be coated with metals or other substances and used in electronics or other industry. One application that has been mentioned is radiofrequency shielding. To keep cell-phones from working inside conference rooms, perhaps?

NaturalNano is not the only entity interested in halloysite nanotubes. Scientists from the Naval Research Laboratory have created a method of controlled release of anti-fouling agents in paint. They hope to use these on U.S. warships to reduce fouling of hulls by barnacles and other sea creatures. NanoDynamics, another small company, has a purchase order in for up to 5000 tons of halloysite clay from Atlas Mines.

Synthetic Nerve Membranes

The nerve cell membrane can be thought of as a naturally designed piece of electronics – which has the advantage of being very thin and very responsive. A small Australian bionanotech company called Ambri is mimicking the nerve cell membrane to create a new sort of biosensor.

The name "Ambri" is an acronym that stands for Australian Membrane Biological Research Institute, which in 1992 was spun out of the largest research organization in Australia, Commonwealth Scientific & Industrial Research Organization (CSIRO). As medical diagnostic companies go, Ambri is still very small, with 75 employees in Australia, and another 18 in that Mecca of biotechnology, Palo

Alto, CA. Their prototype device is a kind of synthetic nerve cell membrane on a chip, which registers the presence of disease markers with a voltage change that can be immediately monitored electronically.

How much better? Ambri's Chief Technology Officer, Bruce Cornell, in an Aussie accent modulated somewhat by a stint in London, describes the sensitivity of the membrane as equivalent to detecting the increase in sugar content of Sydney Harbor after throwing a sugar cube off the ferry. "And it's so simple to use", says Cornell, that "we've even had corporate lawyers working them." How much faster? No more than 5 minutes at the point of care, compared to the 4 to 24 hours it takes to get results from a pathology lab today.

Much of medicine's ability to diagnose disease conditions depends on the recognition of particular biomolecules. Diagnostic tests typically use antibodies that recognize particular disease molecules, or sometimes, hormone receptors which interact with particular hormones overproduced in a disease condition. Oligonucleotides, short chains of specific RNA or DNA sequences, can also be used to recognize their complementary strand.

Ambri's innovation was to combine the recognition properties of biomolecules with the current transducing properties of a nerve cell membrane, to create a new type of biodevice. A nerve cell is an exquisitely complex device that nevertheless relies on simple chemistry – the flow of ions across a membrane – to regulate their response to stimulus. Think of the nerve cell as a kind of capacitor that builds up charge, and then releases it all at once. Only the nerve cell capacitor relies on the flow of positively potassium and sodium ions rather than electrons to fulfill its function. The flow of charged molecules occurs through ion channels formed from proteins.

Ambri's membrane, like that of the nerve cell, is made from two layers of phospholipids; these are negatively charged on the outside and oily in the middle. The membrane is all of 4 nm thick, and sits on a conducting layer, also 4 nm thick, layered onto a gold electrode.

A real biological membrane is incredibly complex, containing not only phospholipids, but also cholesterol, many different types of structural proteins, receptor proteins, and lined on the outside with complex carbohydrates. Ambri's membranes, by contrast, are simple-minded things, just the lipids and a few active proteins – the ion channel itself and some recognition molecules.

Ambri's device received world press coverage in 1998 as the first purpose-built nanomachine operating with moving parts that are only nanometers in dimension. How do you build something like that? Simply put, there are two basic processes involved in its manufacture: thin-layer deposition, which is already used heavily in the semiconductor industry; and self-assembly – the biological components, including the membranes and proteins, assemble themselves, without aid from human hands. Self-assembly is one of the Holy Grails of nanotechnology, but Nature beat us to it by several billion years.

The particular ion channel protein that Ambri uses is called gramicidin. The advantages of this protein are that it is small, very stable, cheap to produce, and it only works when two subunits are in register. The subunits form a bridge across

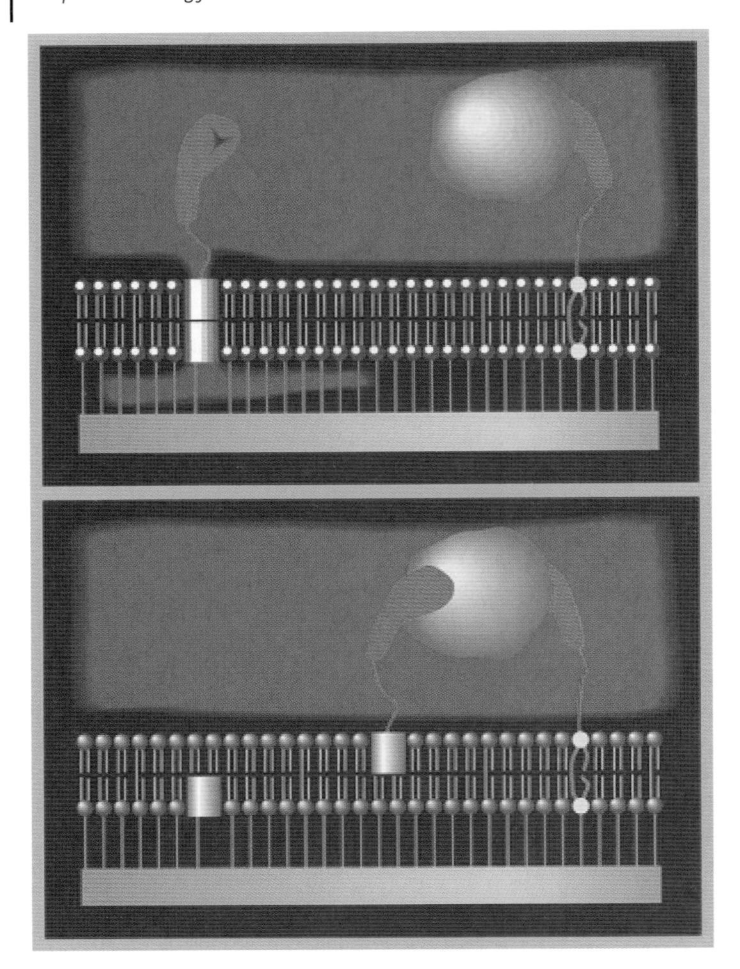

Figure 37 Illustration of Ambri's ion switch. A gold electrode is coated with conducting layer (4 nm diameter). Above these, a bilayer phospholipid membrane is allowed to self-assemble. Floating in the membrane are gramicidin ion channel proteins (pink cylinders) some on the outer membrane, some on the inner membrane facing the gold electrode. When the gramicidin subunits are in register across the membrane, current is flowing as in upper panel. Antibody (red shapes) are attached to the gramicidin subunits. Some antibody molecules are also fixed to a linker that crosses the membrane. An antigen (green globular shape in lower panel) crosslinks the antibody molecules, pulling the antibody together, dissociating the gramicidin subunits and thereby interrupting current flow, as in the lower panel.

the bilayer membrane; in a living cell, one subunit would face out, and one would face in towards the cell. When the two subunits are hooked together, current can flow across the membrane; when they are pulled apart, no current will flow. It is like a molecular switch.

By linking the gramicidin ion channel, which is floating in the membrane with antibodies or other recognition molecules, and by placing similar recognition molecules that are fixed in the membrane, Ambri is able to make a biosensor switch that turns off in the presence of a disease molecule of interest (Fig. 37). The antigen crosslinks one subunit of gramicidin with a fixed antibody, and pulls the two gramicidin molecules apart, breaking the circuit. This is registered as a drop in conductivity across the membrane. Similar tricks can be played using oligonucleotides as the recognition molecules.

This system is easily integrated with electronics; Ambri's Sensidx system is already being tested in hospital laboratories with six different diagnostic tests, including one for glycosylated hemoglobin, an indicator of diabetes, and also in tests for the diagnosis of respiratory diseases. The particular assays under testing were chosen because of "... their attractiveness to potential commercial partners" says Ambri. Until supplies of biological reagents have been secured and freedom to operate cleared from an intellectual property standpoint, the company is not prepared to say anything more about the assays.

Co-Opting Biology

The fastest road to a workable nanotechnology is to start with a nanotechnology that already works – biology. Some researchers have gone beyond trying to mimic Mother Nature and are trying to put the grand old lady to work on their own schemes.

Researchers at the University of Texas, led by Andrea Belcher (now at MIT), have engineered a bacteriophage to contain a peptide with an affinity for zinc sulfide nano crystals and showed that, under the right conditions, a mixture of virus and crystals will build itself into a liquid crystal film. Their work was reported in a paper entitled "Ordering of Quantum Dots using Genetically Engineered Viruses," in the May 3rd, 2002 issue of *Science*.

Liquid crystals are long chains of molecules that uniformly line up to form crystal-like structures under the influence of an electric field. The liquid crystals commonly used in computer displays, for instance, shift their orientation in response to changes in the surrounding electric field in order to change the color of individual pixels on a screen.

It should be possible to use viruses to build many other types of useful materials, according to Belcher, including semiconductor, magnetic, optical, and biocompatible materials. The difficult part of the process is finding particular proteins that can bind and assemble materials like semiconductors. Belcher's group started with a library consisting of about one billion different proteins.

Once the researchers found the protein with the characteristics needed to bind to zinc sulfate, DNA that encoded the protein was inserted into a virus' genetic material so that the protein would be expressed in its coat. Once they had the requisite phage, they could prepare billions of copies by infecting bacteria in the laboratory. The researchers then made a virus and zinc sulfide liquid crystal suspension, and within a week the material had assembled itself into uniform films. The films are ordered at the nanometer scale, extend to several centimeters, and are stable enough to be picked up by forceps.

The virus has dimensions of 6.6 × 880 nm, a long narrow shape. The zinc sulfide nanoparticles are 3 nm in diameter. The peptide that binds to the zinc sulfide nano particles is about 10 nm long. Thus, 1 cm^2 of film contains about 40 billion viruses.

"The exact structure of the film depends on the concentration of the viruses in the liquid crystal suspension and the strength of the surrounding magnetic field," said Belcher. The proteins that make up the virus' outer coat are weakly magnetic, which causes a growing virus-nanoparticle complex to align with a magnetic field. One type of film that the researchers made was ordered into domains, or patterns that spanned 0.07 mm and repeated continuously. Materials with such small-scale patterns could be used to make storage devices. Belcher's group is also developing biotechnology applications.

A few practical applications are possible within five years, but most applications will take 10 years or longer to develop, she says.

An offshoot of Belcher's work has been the establishment of a nanotech company, Massachusetts-based Cambrios, which seeks to use some of her ideas in the semiconductor industry.

Molecular motors abound in the biological world, from the machines that pack DNA into viruses, the ribosome that makes proteins using a ratcheting RNA molecule as source code, to the kinesin molecules that shuttle molecules to and from the cell surface. Such molecular motors might be incorporated into medical devices of the future, for use as drug pumps, etc. Proteins such as enzymes are motors of a limited kind, in that they have moving parts, and can translocate substrates, like a robot on an assembly line.

Cornell biological engineers Carlo Montemagno and George Bachand devised the first self-propelled nanobot. The Cornell University scientists genetically tweaked an F1-ATPase protein, and then attached one end of their nanomachine to a metallic substrate and affixed a tiny nickel rod as a propeller to the other. As the ATP began to break down, the bio-motor started to move, and ran for 40 minutes before it was shut off. Once the engine is outfitted with a compartment for storing antibiotics, the device could be turned into a drug pump capable of entering individual cells, according to Montemagno. The entire device, including the motor and propeller on a nickel post, was comparable in size to some virus particles.

Bachand has joined Sandia, becoming that national laboratory's first molecular biologist. He plans to continue his efforts in using living motor proteins to power nanoelectromechanical (NEMS) systems, along with other projects.

Various researchers have made motors out of DNA molecules. Nadrian Seeman's group at New York State University has created a motor out of a four-stranded DNA molecule that will go through a mechanical cycle over and over again. The motor is power by "fuel strands" of DNA that bind to the motor at various places.

Chapter 7
Nanoelectronics

Nanoelectronics – yet another nanoneologism (whoops!, there goes another one). But in this case, nearly everybody in the electronics industry agrees that there is nowhere the industry can go but down further in scale, as it has been doing since Thomas Edison's day. Semiconductor chips, on a commercial scale, have already breached the 100- nm feature size, which puts it in the realm of nanotechnology. Several nanolithography techniques promise to push that limit down to 10 nm, which is so small that you could count the individual atoms in cross-section without making it into triple digits. IBM is already working on memory devices that will use individual atoms to represent computer code. Eventually, it is expected that electronic components: switches, transistors, capacitors, resistors, actuators, attenuators, potentiometers, and the like, will be made up of single molecules, connected by circuit wires that may also be single molecules. This field, called molecular electronics is still a little too far out there to be discussed intelligently by the non-cognoscenti. When it does come, it will make the wiring of the human brain look positively primitive.

We will concentrate in this chapter on developments that have already occurred, or which are visible on the horizon.

Spintronics

Rumpelstiltskin, according to the Brothers Grimm, was a weird little man who made a name for himself by his ability to spin straw into precious gold. The ancient alchemists, from time immemorial up to the birth of science, wasted much effort in trying to convert lead into gold. Sir Isaac Newton, for instance, is said by some, to be not only the first real scientist, but also the last alchemist. James Daughton, on the other hand, is a very bright guy who hopes to convert the natural spin of electrons into gold, or at least dollars, for the company he founded, Non-Volatile Electronics Corporation, now known as NVE Corp. Well-behaved electrons, it turns out, are worth a whole more by weight, and are generally more useful than gold.

The Nanotech Pioneers. Steven A. Edwards

ISBN: 3-527-31290-0

Daughton started NVE in 1989 after fifteen years at Honeywell, where he was a vice president managing solid-state electronics R & D. By 1994, NVE had already commercialized its first magnetic sensor based on a newly discovered property of matter, giant magneto resistance, a consequence of electronic conductivity based on spin. With that product, commercial spin electronics, or spintronics was born.

"The term 'spintronics' evolved during a period of stunning discoveries and developments over the past 15 years regarding magnetic phenomena in the areas of magnetoresistance, magnetism switching, and other magnetic properties," says Daughton. Credit for coining the word "spintronics" goes to Stuart Wolfe of the Defense Advanced Research Projects Agency (DARPA), one of the spookiest of government agencies. DARPA has been intimately involved with funding the development of this nascent industry.

Ever since electronics began in earnest over a century ago, when Thomas Edison created the electric light, the industry has relied on the fact that negatively charged electrons flow naturally through a metallic conductor toward a positively charged pole. The entire electronics industry – lighting, heating, refrigeration, television, computers, MP3 players, digital cameras – and, as a consequence, most of our industrialized society depends on this property of electrons.

Among the most important products that the electronics industry has given us are semiconductor chips that allow us to manipulate and store information. Every year, these become smaller and more powerful, in accordance with Moore's law. However, as the feature sizes of these chips sink into the nanoscale realm, their designers must deal with the non-intuitive effects of matter described by quantum physics. Some fear that we are fast approaching a physical limit beyond which computer chips cannot go. Others, like Daughton, see in quantum physics the road to a whole new class of products that use quantum effects to their advantage.

Conventional electronics relies on charge, but electrons have at least one other trick, a somewhat mystical quantum property called spin. So what is spin exactly? It is probably easiest to think of each electron as a tiny magnet. The macroscopic magnets with which we are familiar have two poles, north and south. Electrons likewise have two poles, which physicists have helpfully labeled "up" and "down." This magnetic property of electrons is related to spin. Think of a big charged sphere rotating in space (the Earth, for instance). The spinning electronic field generates a magnetic field with two poles. Unlike the Earth, however, a single electron only has one pole (it is hard to imagine, after all, both a north and south pole co-habiting on an electron, which is virtually a dimensionless point). In fact natural magnets are generated when domains of spin-oriented electrons line up, with the "up" electrons on the north pole and the "down" electrons collecting on the south pole (Fig. 38).

Spin is also sometimes described as an "angular momentum" which, in the macroscopic world, is the momentum of a spinning body like a planet or a baseball. A baseball loses its angular momentum when it stops spinning; for an electron, spin is an intrinsic property and never goes away. It is probably not accurate to think of an electron as a little spinning ball, but our physical intuition fails us when we get down to the quantum-scaled universe.

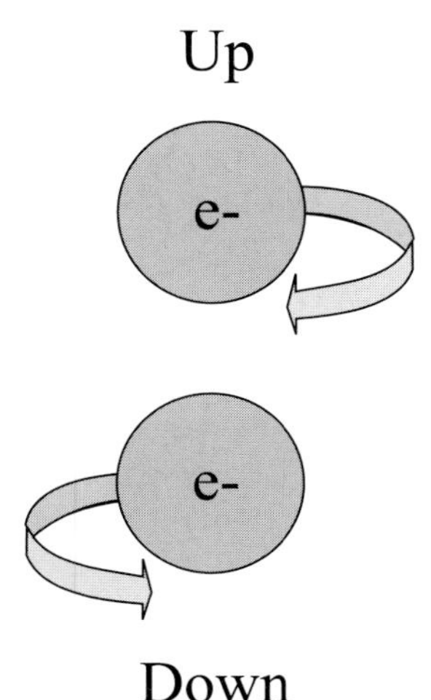

Figure 38 Electron spin. Electrons have two spins called "up" and "down." This property is related to magnetism – natural magnets occur when electrons with an up spin collect on the north pole and electrons with a down spin collect on the south.

The first major products to arise from spintronics are the read-heads on the latest generation of hard drives. These rely on the spintronic property called "giant magneto resistance." Certain materials change their resistance to electronic current depending upon their orientation in a magnetic field. Read-heads rely on nanotech techniques for creating very thin films. A sandwich is made in which two magnetized layer are separated by a very thin, non-magnetized layer. The magnetized layers will allow only electrons with a certain spin (either up or down) to pass through. If they are aligned, there will be low resistance, allowing current to pass through. If they are not aligned, resistance is created such that electrons of neither spin will pass through. An analogy can be made to a polarized light filter. One such filter will pass only light coming from a perpendicular angle, thus eliminating glare. But if two filters are placed at right-angles to each other, almost no light will pass through.

In the read-head, the first magnetic layer is fixed in one direction, but the second is not. As it passes above the track of data on a hard drive, the little magnetized domains that represent the 1s and 0s of computer code flip the second layer from parallel to antiparallel, changing the resistance, and thus the current through the read-head. The read-head flies at up to 80 miles per hour over the surface of the disk at a vertical separation of only 10 nm, reading or writing as it goes. Current computer hard drives are magnificent examples of nanoscale engineering, although largely unrecognized as such.

NVE has patents and licensed intellectual property encompassing giant magnetoresistors as wells as a second important nanotech device, spin-dependent tunnel junctions. The latter device also employs thin films, as small as a few atoms thick. It allows the resistance in an insulating layer to change as the result of the spin-alignment in an adjoining layer. This effectively functions as a switch that creates an interface between spintronics and conventional electronics.

To help turn its patents into products, NVE brought Daniel Baker on board in 2001, with Daughton relinquishing his title of CEO so that he could concentrate on advancing the technology. Baker was formerly President and CEO of Printware, Inc., which makes high-speed laser printing systems.

NVE currently has an interest in three types of products: sensors, couplers and MRAM, a new type of computer memory.

NVE's sensor products detect the position of a magnet or of a metal to determine position or speed. Their biggest market for such sensors is in industrial robots. The sensors are also used in implantable medical devices; St. Jude Medical is one of their biggest customers. NVE is also targeting non-life-support medical devices such as hearing aids, which have a shorter FDA approval cycle. Additionally, NVE is working on what it calls "BioMagnetIC" sensors, after the DARPA project of the same name. NVE has announced a contract to tailor devices for a lab-on-chip applications – "ultra-compact, fast, accurate diagnostic systems to replace entire laboratories," according to Baker.

NVE's couplers are a component of transceivers used to transmit data. NVE's spintronic couplers operate at 110 million bits per second, better than twice as fast as the best optical couplers. So far, NVE's couplers are mainly used in for factory and industrial uses. According to Baker, NVE also has "several broadband and telecommunication design wins, although the volumes aren't very large yet."

But the biggest application for spintronics in the near future is MRAM (for magnetic random access memory). There are currently three major types of memory chips in the computing and communication industry: DRAM, SRAM, and flash memory. DRAM, or dynamic access memory, is the type of chip used to operate computers and accounts for the biggest market, about $24 billion in 2004. SRAM, or static random access memory, is a conventional memory chip that is faster than DRAM but lower in density; it is used for operations such as digital signal processing in cell phones and caches in computers, where speed is critical (see Table 9). This was about a $3 billion dollar market in 2004. Flash memory is non-volatile, which means it doesn't go away when the power goes off. Flash memory is used in cell phones to store data and in memory sticks, personal digital assistants (PDAs), MP3 players and digital cameras. This was a $12 billion market in 2004, and is growing rapidly.

MRAM has the speed of SRAM, the density of DRAM, and the non-volatility of flash memory. Could MRAM potentially replace all three markets, worth many tens of billions of dollars? Says Baker, "That's our vision if we can continue to improve density, cost, and power consumption. We have several programs underway to make that happen, with patents in each area: 1) High-sensitivity spin-dependent tunnel junctions – New spintronic recipes that can reduce write current that also allows smaller thermally limited cells; 2) vertical MRAM and spin-momentum writing – these designs and inventions can shrink cell size and reduce write current.; and 3) magneto-thermal MRAM – an invention that uses a combination of magnetic fields and ultra-fast heating from current pulses to reduce the energy required to write data." Magnetothermal RAM actually uses waste heat generated on the chip to help speed changes of change of magnetic state, a way to turn a long-standing problem – heat generated by computer chips – into an advantage.

Table 9 Types of memory chips.

Type of memory	Abbreviation	Most important feature	Used for:
Dynamic Random Access Memory	DRAM	High density	Computer operating memory
Static Random Access Memory	SRAM	High speed	Cell phones, computer caches
Flash Memory	–	Non-volatility	PDAs, cameras
Magnetic Random Access Memory	MRAM	High density, speed, non-volatility	All uses
Nanotube Random Access Memory	NRAM	High density, speed, and non-volatility	All uses

In an MRAM chip, data are stored in the spin of electrons in thin metal films, and are read with NVE's spin-dependent tunnel junctions. Unlike the DRAM chip that provides the working memory for your computer, MRAM is essentially permanent, like your hard drive. Which means that computers of the future will be booted instantaneously; the hard drive will become superfluous. In the closer-in future, MRAM chips may also replace battery-powered SRAMS for factory control, point-of-sale terminals, gaming electronics and military electronics. MRAM should find application in cell phones, PDAs, and digital cameras.

NVE licenses its intellectual property relative to MRAM to several companies including Motorola (and through Motorola to Freescale Semiconductor), Cypress Semiconductor, Honeywell, and Union Semiconductor. Motorola and Cypress both own equity positions in NVE. Freescale has produced modest quantities of an MRAM chip, which it is now sampling to customers, and hopes to have full-scale production in 2005. Cypress is behind its own announced schedule in producing MRAM chips.

Unfortunately, NVE's licensees are not the only ones who want to develop MRAM chips. The competitors include a Who's Who of semiconductor manufacturers, including Fujitsu, Hewlett Packard, IBM, Infineon, NEC, Samsung, Sony, Taiwan, and Toshiba. For NVE, these represent a daunting competitive landscape. On the other hand, with this many heavyweights committed to MRAM, the question is not whether MRAM will be adopted generally, but when, and who will profit.

NVE does not have the resources to build its own MRAM fabrication plant. "Because of the large investment required for a memory factory (several billion dollars) our strategy has been to license our technology to memory manufacturers," explains Baker. "Therefore we do not plan to build MRAM ourselves, but acquiring additional licensees is certainly part of our strategy."

MRAM technology itself is not the only new technology out there: there is also NRAM, for Nanotube Random Access Memory. This technology, which is being promoted by the start-up nanotech company Nantero, uses carbon nanotubes as individual switches to record the 1s and 0s of computer memory. According to Nantero's CEO Greg Schmergel, "NRAM is as dense as DRAM, as fast as SRAM, and as non-volatile as flash memory." Sound familiar? Nantero is targeting basically the whole computer memory market, which it estimates at about $100 billion annually.

"From what I know... [NRAM] is interesting technology, but we that believe MRAM is closer to commercialization, more producible, and more scalable than carbon nanotube memory," says Baker.

Nantero does have some adherents, however. LSI Logic is seeking to integrate the NRAM concept into its CMOS technology. NRAM chips started rolling off LSI's production lines in May 2004, converting its engineers into nanotech believers. BAE Systems is working with Nantero to develop NRAM chips for defense and aerospace application.

Looking farther forward, what lies beyond MRAM for spintronics? "The short answer is 'I don't know'," admits James Daughton, "... but there are plenty of possibilities. There is a lot of activity in spins interacting with light. Spin currents may be generated in semiconductor materials without the use of ferromagnetic materials. Also, very high frequency signals are being generated in ferromagnetic materials by injecting spin polarized currents into them, and there is a lot of research on ferromagnetic semiconductors. Spintronic devices have been proposed for logic and quantum computing."

Like NVE, IBM is working on spintronic devices, including MRAM. The difference of course is the level of resources; IBM's research budget dwarfs even the U.S. governments Nanotechnology Initiative. IBM helped launch the first mass-produced spintronic device for the hard-disk drive industry; introduced in 1997, the giant magnetoresistive (GMR) read head was developed at the IBM Almaden lab. The magnetic-field sensor that allowed data density rates to increase by 40 times their previous levels. IBM is collaborating with Stanford University: the partners have said that their research in the field of spintronics could lead to reconfigurable logic devices, room-temperature superconductors and possible even quantum computers. According to Robert Morris, director of the Almaden lab, spintronics breakthroughs could revolutionize the electronics industry in the next few years as much as the transistor did fifty years ago.

Scientists at the College of Nanoscale Science and Engineering (CNSE) at the University at Albany have published research that could lay the foundation for using silicon to develop chips with magnetic properties, possibly speeding the development of spintronic devices. Silicon and other semiconductor materials are used to make chips for random access memory and central processing units. Recent research has discovered that a semiconductor can be made magnetic by doping it with an impurity such as manganese (Mn). The resulting material, called a diluted magnetic semiconductor (DMS), combines the properties of magnetism used in permanent information storage with that of semiconductor memory and logic devices. DMS spintronic devices have the potential to operate at considerably higher speeds and consume less power than conventional devices.

CNSE Professor Vincent LaBella and Martin Bolduc, CNSE post-doctorate fellow, showed for the first time that silicon can be made "ferromagnetic" or permanently magnetic at normal operating temperatures for computer devices. The researchers achieved this by implanting manganese into silicon up to a concentration of 1 % per atom. They found the silicon was ferromagnetic above room temperature, or up to 127 °C.

If spin-polarized currents could flow in semiconductors instead of metals, this would allow many more types of spintronic devices that could use the semiconductors' high-quality optical properties as well as their ability to amplify both optical and electrical signals. Examples include ultrafast switches and fully programmable all-spintronics microprocessors. Your PDA or cell phone could then employ very rapid spintronic devices.

In 1990, Supriyo Datta and Biswajit A. Das, who were then working at Purdue University, proposed a design for a spin-polarized field-effect transistor, or SpinFET for short. So far no one has been able to build the SpinFET, but work on doping semiconductors with metals is moving us closer.

In a conventional FET, a narrow channel composed of a semiconducting material runs between two electrodes called the source and the drain. When voltage is applied to the gate electrode, the resulting electric field drives electrons out of the channel, turning the channel into an insulator. The SpinFET would have a ferromagnetic source and drain so that the current flowing into the channel is spin-polarized. When a voltage is applied to the gate electrode, the spins rotate as they pass through the channel and the drain rejects these anti-aligned electrons, making the channel effectively an insulator.

Flipping an electron's spin can be done faster and with a lot less energy than it takes to force an electron out of a channel. One can also imagine introducing additional levels of control that would allow logic gates to be altered on the fly.

The spin of the electron has also been proposed as the most likely candidate for the "qubit" – the fundamental unit of quantum computing, like the coded 1s and 0s of conventional computing, only taking up a lot less room and more easily addressable. Quantum computing will be discussed in a separate section in Chapter 10.

Other new spintronic devices may take advantage of the "spin Hall effect" demonstrated recently by David Awschalom, a professor of physics and electrical and computer engineering at the University of California, Santa Barbara. The classical Hall effect is named after American physicist Edwin Hall, who discovered it in 1879. This effect occurs when an electric current flows through a conductor within a magnetic field. The magnetic field exerts a force on the moving charge carriers, pushing them to one side of the conductor. The resulting build-up of charge at the sides of the conductor ultimately balances the force of the magnetic field – the result is a measurable voltage between opposite sides of the conductor. This effect is used to good advantage by today's sensors and electronics.

A "spin Hall" effect was predicted in 1971 by two Russian physicists, M.I. D'yakonov and V.I. Perel. By analogy to the classical Hall effect, current-carrying electrons with opposite spins are predicted to move toward opposite sides of a semiconductor wire, even in the absence of a magnetic field. In theory, this would result in an accumulation of spins at the edges of the conductor with opposing spin polarization – in other words, a "spin current" transverse to the electric field rather than a charge-dependent voltage drop. Because no net charge is flowing, the "spin Hall" current has been difficult to observe. Awschalom accomplished the trick with a high-resolution Kerr microscope, a tool for imaging the magnetic microstructure of a sample.

"The most exciting aspect of this finding is that you don't know exactly where it's going to lead," said Awschalom. Potential uses in sensing devices in optical communication are possible.

When it comes to electronics, spin offers a completely different functionality than charge. In his classic essay, "There's Plenty of Room at the Bottom," Feynman challenged us to build electronic circuits based on the "interaction of quantized spins." In addition to memory devices, spin-based transistors, LEDs and lasers are already being proposed. The greatest applications for spintronics are not those analogous to conventional electronics, believes Awschalom, but those we haven't even imagined yet.

Nanotube Memory Chips: NRAM

Nantero, a tiny entrepreneurial company, has the modest ambition of replacing all existing computer memory chips (not to mention those in cell phones, digital cameras, and MP3 players) with its carbon nanotube-based NRAM chips, a market estimated at around $100 billion. According to CEO Greg Schmergel, the advantages of NRAM are speed, density, non-volatility, lower power consumption, and reduction of errors due to alpha radiation.

Nantero has raised $16.5 million in two rounds of financing, and has attracted some impressive industry leaders to its board. Among these are Alex D'Arbeloff, cofounder of Teradyne, and now chairman of MIT, and former IBM executive O.B. Bilous, who is now chairman of the board for International SEMATECH, a global consortium of semiconductor manufactures.

Nantero has already worked some deals with some major players in the semiconductor industry. LSI Logic is integrating the NRAM concept with its CMOS manufacturing technology. LSI will have the option of using NRAM to replace S-RAM in its ASIC (application-specific integrated circuit) chips. An earlier collaboration with ASML Holdings, a semiconductor equipment company, demonstrated that carbon nanotubes were compatible with ASML's lithography systems.

Nantero and BAE Systems will collaborate to develop carbon nanotube devices for aerospace and defense applications. Schmergel suspects that NRAM memory might be stable in the face of an electromagnetic pulse engendered by an atomic blast, but the company has yet to carry out such experiments.

Like NVE Corp., Nantero has no intentions of building a fabrication plant and making its own chips. Instead, the company will rely on royalties from its intellectual property, a business model for the semiconductor industry already pioneered successfully by Rambus and Qualcomm. Schmergel expects the first commercial quantity NRAM chips to be marketed in two to three years.

Nanowires

The nabob of Nanosys, which is the nexus of nanowires, the neatest nascent nanotech around, is none other than Larry Bock, previously founder or co-founder of Neurocrine Biosciences, Athena Neurosciences, Argonaut Technologies, Onyx Pharmaceuticals, Genpharm International, Caliper Technologies, and Illumina, not to mention Pharmacopoeia, Vertex Pharmaceuticals, and Ariad Pharmaceuticals. All those other companies are biotechnology companies, some of the most successful biotechnology companies around. What does Nanosys have to do with biotechnology? Well, nothing, nada, ningun, nyet, *per se.*

Larry Bock's undergraduate experience occurred in a small liberal arts type of place, Bowdoin College. He went on to receive a master's degree in business administration from UCLA. Though these academic achievements are not to be sneered at, you might think that they wouldn't give him a whole lot of background to enter the high-tech arena. You might even be correct about that. But what it did prepare him for was venture capital as a business. Bock is the most serial of serial entrepreneurs. Biotech was where it was; nanotech is now where it's at. If you need confirmation of that, look to see where Larry Bock is at.

Nanosys has on its scientific advisory board some of the most outstanding academic nanoscientists around, including Louis Brus, Charles Lieber, Paul Alivisatos, Moungi Bawendi, James Heath, Hongkun Park, and Peidong Yang. The company has spent, reportedly, upwards of $40 million on intellectual property, much of it from the laboratories of the people just mentioned.

If you look at the Nanosys website, they talk of applications that websurfers can understand – solar power cells, flexible electronics, fuel-cells and memory devices. But look at where their intellectual property investments are going, it points to two things: zero-dimensional and one-dimensional inorganic electronic devices, also known as quantum dots and nanowires, respectively. Louis Brus, Paul Alivsatos, and Moungi Bawendi are innovators in the field of quantum dots, which we have covered previously, in Chapter 4, Nanoparticles and other Nanomaterials. Charles Lieber, Piedong Yang, James Heath, and Hongkun Park are illuminati in the field of nanowires and nanoelectronics.

"The work I am doing and consider to be most innovative and of greatest long-term significance," said Charles Lieber, at the 2004 U.S. Technology Awards ceremony, "centers on advancing the bottom-up paradigm for nanotechnology through the synthesis and understanding of fundamental and unique properties of nanoscale wires, the development of approaches for assembly of these nanoscale wires into integrated structures, and the definition of applications of these nanoscale materials in technologies ranging from nanoelectronics to biosensing and photonics."

Nanowires are more or less like ordinary wires, only a lot thinner. How thin? About 10 atoms in diameter is state of the art. The length can be a lot longer – up to 1000 nm (1 μm) can be achieved (Fig. 39). Unlike copper wires, which result from the purification of bulk copper that is then drawn out into wires, nanowires are chemically synthesized from the bottom-up, molecule by molecule. You might also characterize it as crystal growth, but constrained to one dimension.

Recently, Nanosys announced the issuance of U.S. Patent No. 6,882,051 titled "Nanowires, nanostructures and devices fabricated therefrom." The patent, licensed from the Regents of the University of California, covers "fundamental compositions of matter and methods for creating novel nanowire heterostructures in which the composition changes longitudinally along a wire's length and/or coaxially about its width." Piedong Yang is one of the inventors named.

The technology covers a whole host of devices that might be constructed from such wires, including field effect transistors (FET), Light Emitting Diodes (LEDs), nanolasers, solar cells, thermoelectric devices, optical detectors, and chemical and biological sensors. A broad patent like this is part of the land-grab that is going on in nanotechnology today. The major claims of this patent seems to relate to anything that can be made from any nanowire composed of a "first segment of a first material; and a second segment of a second material joined to said first segment..."; in other words, a nanowire composed of alternating segments of two different materials. The press release announcing the patent gives the example of a nanowire composed of silicon and silicon germanium, but the patent is not nearly so specific.

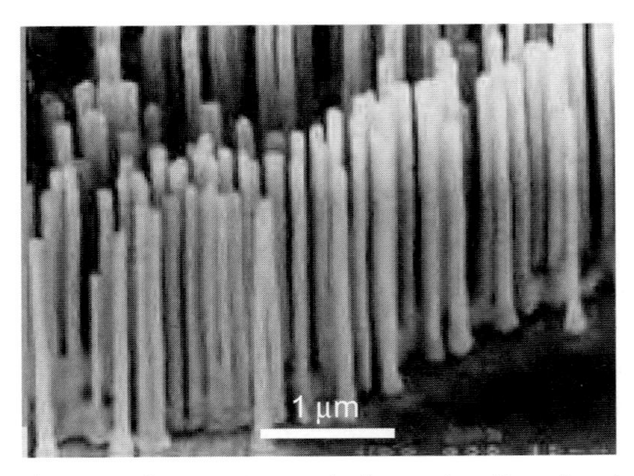

Figure 39 Electron micrograph of nanowires. Reproduced courtesy of Piedong Yang, Professor of Chemistry, University of California, Berkley.

By making a nanoscale wire from two different materials which may have vastly different properties, you take it out of the intuitive macroscale concept of "wire" into something that is a whole lot more like a "device" – something that does not just carry current, but can alter the information carried by the current in various ways.

Nanosys does not yet build or market anything, nor does it yet collect royalties on anything that is built or marketed. But one gets the idea that, in fairly short order, they will.

In 2004, the nanotech world held its breath when Nanosys announced plans for an initial public offering (IPO) of its stock. Although not the first nanotech company to go public, it was the first widely recognized company to do so after nanotech had become well-known to the Wall Street community. I had made one of Nanosys's officers a keynote speaker at conference hosted by Business Communications Company in May of that year. SEC rules prohibit public transmission of information about a company except through its prospectus during the "quiet period" prior to an IPO. Naturally, my speaker had to withdraw on the order of the company's lawyers. The Nanosys quiet period lasted through the remainder of 2004 and into 2005. Market conditions for IPOs were, to say the least, terrible. If Nanosys wouldn't brave it, with the credibility provided by Larry Bock and its stable of stellar scientific advisors, it was unlikely that any other nano company would, and for the most part, this appears to have been the case. Cambridge Display Technologies (see below) was an exception – its IPO was well received in December 2004. Unlike Nanosys, Cambridge had the advantage of product-based royalties, although the company is still a long way from profits.

For now, it appears that Nanosys has given up its plans for an IPO. In the meantime, the company has announced a series of government contracts and co-development agreements with major companies. The IPO, when it does come, is still likely to be a rich reward for the Nanosys' venture investors.

Thin Films of Glowing Polymers

One day in 1989, not long after submitting his Ph.D. thesis, Jeremy Burroughes found something that he wasn't looking for – an incidental observation that may well have launched an industry. Working in the same storied Cavendish laboratory where J.J. Thompson discovered the electron, Burroughes was carrying out some exploratory studies with organic polymers, essentially a type of plastic. While passing electricity through the polymers, Burroughes noticed that they began to glow. It was just a little glimmer of light, dim and of low efficiency, but Burroughes' mentor, Richard Friend, immediately realized the commercial potential of the discovery. A paper in *Nature* on "organic electroluminescence" and a patent quickly followed. Two years later, prototype displays containing all of 15 pixels were made using the glowing polymers. The year after that, Cambridge Display Technology (CDT) was born. Today, Jeremy Burroughes is the Chief Technology Officer for that company.

Cambridge University, from which CDT got its name, was already an old institution when Isaac Newton attended classes there as a lad some 300 years ago. Legend has it that he was an abysmal student, but his later accomplishments were enough to get him appointed Lucasian Professor of Mathematics (the same post that Stephen Hawking holds today). Not only did Sir Isaac invent calculus (the bane of my first year in college), but he was also apparently the first person to notice gravity. Reportedly, Newton held up publication of his calculation for the acceleration of gravity because he was embarrassed that he could only describe gravity mathematically without being able to explain what caused it to work. Fortunately for Newton, academic science was not nearly as cut-throat in his time as it is today, so he was still able to claim priority for his gravitational theories when he finally got around to submitting his results. Newton would be astonished to learn that we are no closer today than he was then in understanding why exactly it is that we don't just float off into space. We should all be embarrassed.

Burroughes, Friend and their colleagues at Cambridge were not in possession of a completely satisfactory theory on how or why organic polymers glowed when they were supplied with current, but nevertheless they were not shy about pushing ahead with possible applications. It used to be that science preceded technology. Now, they inform each other in feedback fashion to create a virtuous spiral upward.

Amazingly, CDT is the first company spun out of the hallowed halls of Cambridge University ever to go public. CDT made an initial public offering on NASDAQ in December of 2004, in the midst of what was otherwise a dismal IPO season.

What can Cambridge do with a glowing polymer? Well, for one thing, they can make LEDs – hence the company's ticker symbol, OLED, for organic LED. Cambridge is not the only company seeking to commercialize OLEDs; no less a company than Kodak already has its OLEDs in production for displays on consumer products like cell phones. Kodak, however, uses small molecules (monomers) as the basis for its products; CDT's innovation was to use polymers instead; we will call them PLEDs (aka P-OLEDs) in order to distinguish them.

A PLED is a kind of sandwich. One piece of bread is a transparent electrode, typically made of indium tin oxide. The other piece of bread is a metallic electrode; calcium is often used (Fig. 40). The peanut butter and jelly is composed of organic polymer and is usually no more than 100 nm thick (or one-tenth of one millionth of a meter) – too small to be seen in cross-section, but not too small to emit light. Electrons flow from one electrode to the other, lighting up the polymer that is in-between. The whole thing is layered on to a substrate, which can be glass or plastic, so PLED displays could be made that bend around corners.

Now, you say, it can't be that easy to spread materials in consistent layers only 100 nm thick. Actually, there is a very easy way to do it for polymers, if not for peanut butter. You use an ink-jet printer. How simple is that? Therein lies the difference between polymer and monomer organic LED production. The latter cannot easily be processed from solution, making manufacturing processes very complicated. PLEDs can be molded into any pixel size or shape and can, in principle, be made as high in resolution as the eye can distinguish.

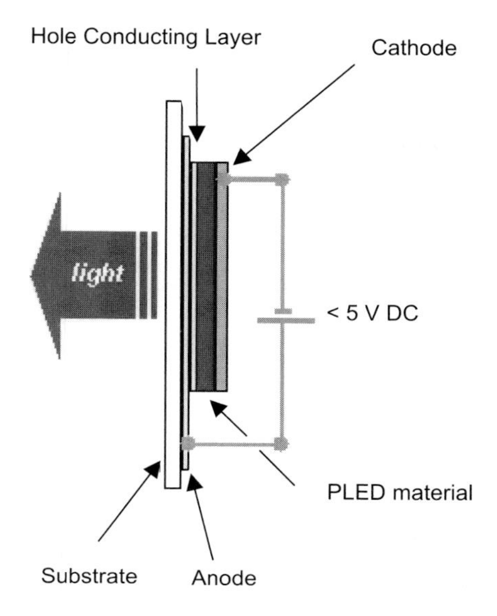

Figure 40 Diagram of a polymer light-emitting diode (PLED). Reproduced courtesy of Cambridge Display Technologies, Ltd., Cambridge, UK.

So what is Cambridge doing with its PLEDs? Terry Nicklin, the marketing director of CDT helped me out with this question. Already, Cambridge PLEDs are being used in low-resolution displays on such consumer products as cell phones and MP3 players. Philips makes a flashy-looking men's shaver featuring a PLED display that appeared in the James Bond film *Die Another Day*. Point-of-purchase displays are another big opportunity. Terry offers as a possibility a DVD case bearing a painted-on PLED display that would show cuts from the film inside to the interested consumer. Electronic shelf-edge displays are another possibility. If Wal-Mart had all its shelf-edges electronically programmed with electronic price displays, then a functionary sitting in his office in Arkansas could change prices for a product in hundreds of stores all over the world automatically. But the real world current killer application for PLEDs, as of this moment, is the essential replacement of all liquid crystal displays.

Nicklin raps off the advantages of PLEDs over liquid crystal displays with a practiced efficiency that suggests endless repetition:

- PLEDs react one-thousand times faster than LCDs.
- PLEDs are brighter.
- PLEDs are thinner and more light-weight than LCDs.
- PLEDs allow 180-degree viewing radius.
- PLEDs have about a four-fold lower power consumption.
- PLEDs are an emissive technology; therefore no backlighting, filters, or polarizers are required.
- PLED displays are much cheaper to manufacture than LCD displays.

Today, LCDs are typically used in computer monitors, television sets, watches, and instrument displays. Ten years from now, LCDs may be long gone and PLEDS will dominate the market. Already, Epson has demonstrated a prototype 40-inch color monitor, using PLEDS. CDT licensee MicroEmissive displays has developed a PLED display for a digital camera that is about to be launched commercially. Nicklin imagines the day when programmable, wall-sized, flat-panel PLED displays will be ubiquitous in markets or airports or anywhere in the world that information needs to be conveyed visually.

PLEDs can do other tricks as well. For instance, PLEDs are reversible; instead of emitting light in response to electricity, they can be made to generate electric current in response to light. Therein lies the scientific basis for a cheap, efficient solar panel. CDT has filed several patents in this area based on work done in the Cavendish laboratory.

A polymer-based solar panel would have more or less the same sandwich structure that a PLED has. One marketing idea would be to have printed polymer solar panels use ambient light to power low-resolution PLED displays on packaging. Lighted messages blinking at us from our cereal boxes – just one of many brilliant, but annoying, innovations awaiting us in the future. Nicklin admits that solar panels are not a priority for CDT right now, however. The company prefers to concentrate on the electronic display industry, currently worth about $50 billion.

CDT would love to follow in the footsteps of General Electric, the largest company in the world by market capital. GE got its start in what scientists call a "brute-force experiment" by Thomas Edison. Lacking any theoretical basis on which to make predictions, Edison spent many, many nights pushing electricity through any type of filament he could get his hands on, to find the one that would both shine brightly and last long enough to make incandescent lighting a reality. One class of filaments that Edison did not try was organic polymers, as they had not yet been invented. If he had, he still would have still settled on tungsten, because the efficiency of polymer lighting is not yet competitive with the good old-fashioned kind. But it is an article of faith among experts that it will be, eventually.

Efficiency, in this case, refers to the ratio of light energy emitted over the energy of electricity used as input. In Burroughes original experiments, this was as low as 0.01 % for organic polymers. The optimism of the Cavendish scientists and their faith in their own abilities as chemists led to the breakthroughs that made PLEDs a reality. Current efficiencies are around 3 %, an improvement of several orders of magnitude. Another doubling or two would make PLEDs competitive with phosphorescent room lighting. This year, in fact, quantum efficiencies of 6 % were obtained using red light-emitting PLEDs in a collaborative project involving CDT and partners Philips, Covion, and some academic institutions.

Nicklin thinks that the first commercial PLED lighting products will be accent-type products. Instead of lamps to light your foyer, you may have a glowing wall of PLEDs for instance. The competition will be as much for the lighting effect as opposed to pure efficiency. Imagine a red glowing ceiling in the boudoir with a dimmer switch!

PLEDs are colored lights. If you add red, green, and blue together, however, you can get a serviceable white light. Blue light has been a special problem for LED manufacturers, as the blue pigments just would not hold up long term. CDT seems to have solved this problem, however, as last year they announced a blue PLED with a useful lifetime of 70 000 hours.

Another approach to lighting is to incorporate a special kind of polymer called a dendrimer (see Chapter 5). Whilst most polymers form into long straight chains, dendrimers are branching polymers that form a sort of tree-like structure. As the polymers branch out from a central core they fall back on themselves to form globular structures, a kind of nanoscale particle. CDT is experimenting with linking phosphorescent small molecules into dendrimers to make them glow. These could be used separately or with CDT's other polymers to make specialty LEDs.

Although CDT has had much success in making its polymers glow, how does it makes it cash flow grow? It should be mentioned that CDT manufactures neither polymers nor PLEDs. The company has agreements with chemical manufacturers to make sure that the supply chain for PLEDs is stocked. The company also owns 50 % of Litrex, a company that makes industrial ink jet printers – again to ensure that there are no kinks in the product pipeline. CDT's own products are the technologies that make PLEDs work. It makes its money through license fees, royalties and joint development contracts. According to Nicklin, CDT has no plans to become a manufacturing company, which would, in effect, put it into competition with its customers. CDT is very much an R&D organization. About 75 % of the company's 120 employees work in research, and many of the remainder, like Nicklin, have a technical background.

CDT's licensees and technology partners already include most of the major players in the flat-panel display market, including Sumitomo, Philips, Seiko Epson, Delta Optoelectronics, Toppan Printing, Kolon Industries, and Osram. Merck KgaA, the world's leading supplier of liquid crystals, having read the writing on the wall, recently purchased CDT collaborator and licensee Covion. Delta has just set up Taiwan's first PLED printing facility using CDT's technology. The two companies have collaborated to build a prototype full-color active matrix display. Kolon Industries is looking to establish PLED manufacturing facilities in Korea. Toppan Printing is exploring a method for using roll printing (think wallpaper!) instead of ink-jet printing to make PLED displays. Sumitomo, which owns an equity position in CDT, is working with the company on new phosphorescent materials, possibly including dendrimers, to use in flat-panel displays.

According to market research firm DisplaySearch, sales from organic LED (both polymer and monomer) displays are expected to increase from about $217 million in 2003 to an estimated $3.1 billion in 2008, a compounded annual growth rate of 70 %. Not coincidentally, CDT's own revenues were up 65 % in 2004 over the prior year, excluding Litrex printer sales. Some way to go, little nano firm. Light up the future.

Nanorobotics

To open his presentation, Silvain Martel shows a slide explaining that nanorobots are not necessarily invisible nanoswarms à la Michael Crichton's *Prey*, but can be small, yet visible robots that have the potential to do work at the nanoscale. Their robotic arms, in this case, are specially designed scanning probe microscopes that can not only image individual atoms but also move them around. The robots themselves are not built to the nanoscale, but they are still tiny, at 32 mm in diameter. Small enough to swallow like a pill, if you wanted to. Martel carries one around in his pocket for demonstration purposes.

Martel calls his robots NanoWalkers (Fig. 41). And indeed they are mobile. He shows a video of the little beasts moving around, shuffling their feet at a rate of 4000 nanosteps per second. But for time-lapse photography, that would be way too fast for the eye to see. They seem agitated, like animated robots in early morning cartoons, only much smaller. They are powered through their toes using a "power-floor". This floor is made up of bands with alternating voltage polarities. The dimensions are such that when one leg of the robot is contact with a positive band, another leg will be in contact with a negative band. This way, the diminutive robots don't have to carry batteries.

Martel is a engineer, but he belies the buttoned-down Dilbert stereotype; perhaps it's the French in his French Canadian background. Pacing back and forth and waving his arms about, he communicates passion about what he is doing. As he talks, he emphasizes not his accomplishments, which are considerable, but the monumental difficulties that still lie ahead. Listening to him talk, it is easy to believe that he chose this project not for its eventual benefits, but for the sheer intellectual challenge of it. It's as if he decided to try his hand at the most hellishly impossible engineering task imaginable.

It is possible, as Don Eigler showed in making an IBM logo out of xenon atoms, to place atoms precisely where we want them. This involves a human operator manipulating a scanning probe microscope through teleoperation – controls that translate the human's large-scale movements of a joystick-like device into much, much smaller movement of the scanning probe tip at the nanoscale. However, because of the necessity of a human operator as part of the control loop to move atoms one at a time, this is a very slow process. Apart from the novelty of the thing, there is no way that much useful work can be accomplished this way. "We anticipate that many tasks performed at the nanometer-scale will sooner or later require high-throughput automation," says Martel. Hence, the need for NanoWalkers.

NanoWalkers are not Drexlerian universal assemblers, but they would accomplish much the same purpose – the positional assembly of molecules into macromolecular, possibly even macroscopic structures. In Martel's vision, they would work in teams of thousands, each with its own particular task, like a hive full of honeybees. In order to accomplish their task, they need very precise information as to their location, down to the atomic scale – quite a bit better that you can get with a GPS locator.

Nano Walker

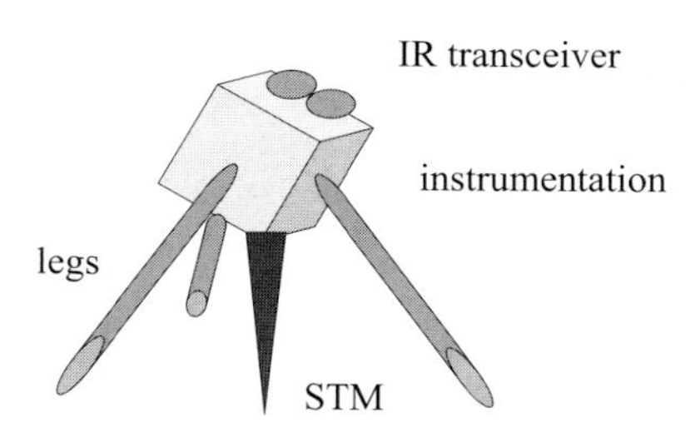

Figure 41 Diagram of a NanoWalker. STM: scanning tunneling microscope.

A central computer mounted on top of the platform coordinates each robot's location and controls them through infrared communication devices built into the robots. Once a robot reaches its designated place on the surface, it begins its assigned task. The machines can be adapted for a number different types of automated research tasks, from measuring material strength to manipulating molecules

Martel is the head of the Nanorobotics Laboratory at d'École Polytechnique de Montréal (EPM), but shuttles back and forth from there to Cambridge, Massachusetts where he works at MIT's BioInstrumentation Laboratory, where much of the work on NanoWalkers is performed.

The test platform for this robotic fleet is a square chamber. The robots – each 32 mm across – are placed inside on a chromium-coated surface that powers the electronics in each NanoWalker.

The prototypes are each rigged with a very small, specially designed scanning tunneling microscope, a device that cannot only create images of individual atoms but also move them. The microscope can make up to 200 000 measurements per second, and the microcomputer embedded in each robot can potentially perform 48 million instructions per second.

Martel envisions one hundred or more robots being deployed on a project, each equipped with a different instrument, working on separate but related tasks. For example, he says they could help to develop new polymers, another project in the BioInstrumentation Lab. While one NanoWalker tests a material's strength, another would measure its optical properties. Both would transmit their results to the central computer, which would incorporate the information and then issue new directions. All without the intervention of a human operator.

Perhaps the future really doesn't need us after all? According to Ian Hunter, head of MIT's BioInstrumentation Lab, the NanoWalker project reflects the lab's search for tools to automate the scientific method itself.

Jim Von Ehr's Zyvex, which bills itself as the first molecular manufacturing company, was founded with the mission of making a practical assembler a possibility. The company already manufactures and sells "nanomanipulators" for use with electron microscopes or scanning. These provide human operators, with their large clumsy fingers, with the ability to move around extremely small objects by remote operation. The company is pragmatically adapting MEMS (microelectromechanical systems) -type technology in the pursuit of the legendary robotic assembler.

The company has been awarded an Advanced Technology grant from the National Institute of Standards and Technology (NIST) to develop prototype microscale assemblers using microelectromechanical systems (MEMS). The company would then extend these capabilities to nanometer geometries, and develop nanoelectromechanical systems (NEMS) for prototype nanoscale assemblers. The eventual aim of the program is to develop systems providing highly parallel microassembly and nanoassembly for real-world, high-volume applications. The Zyvex approach would resemble more an assembly line shrunk to the nanoscale than it would individual autonomous robots.

Chapter 8
Nanotech-Enabled Biomedicine

In the public mind, nanotechnology is all about the mythical tiny submarines that will eventually cruise through our blood vessels, cleaning out clots, attacking bacteria and repairing our body's cells as they go. Innumerable images of these tiny subs already populate the web, illustrating popular science articles on nanotech. Nanobots that cruise through our body doing repair work have already become staple in the science fiction works of authors such as M.M. Buckner or Kathleen Goonan. The apparent genesis of the tiny sub idea was the 1966 sci-fi story, *Fantastic Voyage* by Isaac Aasimov, followed by the movie of the same name, starring Raquel Welch. Ms Welch played a beautiful surgical assistant, who was miniaturized along with her boss, a neurosurgeon, so that they could board a tiny sub and navigate the bloodstream of an important scientist to eliminate a potentially fatal blood clot in his brain.

Futurist Ray Kurzweil even adopted the title "Fantastic Voyage" for his book (Rodale Press, 2004), written with physician Terry Grossman, which claims that nanotechnology will eventually provide humankind with an essentially unlimited lifespan, and that today's baby boom even has a shot at immortality, if only we can get through the intermediate decades before this medical revolution takes hold. "Long-term," said economist John Maynard Keynes, "we are all dead," – but perhaps nanotech-enabled medicine will finally prove him wrong.

Robert Freitas, author of the multivolume series *Nanomedicine*, defines nanomedicine as "the comprehensive monitoring, control, construction, repair, defense and improvement of human biological systems, working from the molecular level, using engineered nanodevices and nanostructures." This seems an adequate definition, for our purposes.

We will not get distracted here with a debate on what constitutes medicine, although in Chapter 9 we will discuss the impact of nanotech on ethical issues in medicine. In centuries past, a sex-change operation would have been regarded by most observers as an abomination rather than a medical procedure. Even today, many question whether such an operation fulfills a legitimate medical need. There is no doubt, however, that a successful sex-change operation requires an experienced team of medical personnel: a surgeon, an endocrinologist, nurses, and perhaps a psychiatrist. Medicine is what doctors do, for good or ill, in our limited view.

The Nanotech Pioneers. Steven A. Edwards

ISBN: 3-527-31290-0

Nanotech detractors like to claim that the real products of nanotech are decades off, as if the mythical tiny submarines and nanobots were the only products that matter. But in the present day, nanoparticles are already being employed in medicine to make the delivery of drugs more efficient. Other nanoparticles are being used to improve the imaging of tissues by MRI and possibly X-rays. Previously, we have described one example of a nanotech diagnostic device by Ambri (Chapter 4), and more are described below. Maybe these relatively prosaic uses of nanotech will prove to be the precursor to the indwelling nanobots of the future.

Developing medical products, both devices and drugs have one enormous hurdle to overcome that other industries do not face. In the U.S., it is called the Food and Drug Administration (the FDA), but every industrialized country has a similar organization. The FDA regulates not only food and drugs, but diagnostic tests used on patients and any kind of medical implant.

Table 10 lists some of the divisions of the FDA concerned just with medical devices, and will give some idea of the Agency's scope and reach.

Table 10 Food and Drug Administration bodies involved in medical device regulation.

Center for Devices and Radiological Health (CDRH)
Office of Device Evaluation
Office of Compliance
Office of Health and Industry Programs
Office of Science and Technology
Office of Surveillance and Biometrics
National Center for Toxicological Research
Center for Scientific Review
Working Group on Review of Bioengineering and Technology and Instrumentation Development Research
Biotechnology Working Group (CDRH)

The Center for Devices and Radiological Health (CDRH) regulates all medical devices sold in the U.S., and monitors their performance. Another important agency is the Office of Compliance, the Division of Bioresearch Monitoring of which develops and monitors surveillance and compliance programs for the medical device industry. The Office also provides guidance to manufacturers in the preparation and submission of regulatory applications. The Office of Science and Technology, among other activities, carries out risk assessments and hazard analyses of devices with respect to tissue interactions and toxicology. The Office of

Health and Industry Programs will work with manufacturers during the design phase of device development to assure compliance with regulatory standards. The Working Groups mentioned in Table 10 evaluate the current state of technology and provide guidance for the formulation of policy to CDRH and the FDA. The InterCenter Tissue Engineering Working Group is tasked with suggesting policy with regard to the rapidly evolving technology of manipulating living tissue.

Drugs must pass a particularly strenuous review process. First, a drug candidate is usually tested on animal models of human diseases. This is partially to determine the efficacy of the drug and partly to weed out any compounds that have unexpected toxicities. After animal testing come human clinical trials, which are divided into Phases (Table 11). The first Phase is mainly to determine safety, and usually involves healthy volunteers. Some chemotherapy drugs, however, are only tested on cancer patients, because toxicity is inherent in their action. When they work, they do so by killing the tumor a little faster than the patient.

Table 11 Human clinical stages of drug candidate review.

Stage	Phase I	Phase II	Phase III	Approval process	Phase IV
Time	1 year	1–2 years	2–3 years	1–2 years	Indefinite
Patients	20–80 healthy volunteers	100–300 patients	1000–3000 patients	Population of drug users	
Purpose	Safety and dosage determination	Safety and efficacy	Efficacy, adverse reactions	FDA Review	Post-marketing monitoring of side effects

A second Phase (II) involving a larger number of patients tests for the safety of the drug, and is also used to determine proper dosage. Finally, Phase III can involve thousands of patients. This phase determines efficacy, and hopefully will discover any rare toxic reactions. Once the clinical testing is done, it can still take the FDA one to two years to make up its mind about approval. This time lag is really due to the under-funding and under-staffing of the agency. Rapidly advancing technology puts additional stresses on the agency as it is difficult to find competent people who are current with the technology.

The requirement for extensive testing makes the process of drug testing very expensive – $800 million per new drug is the figure that is usually given. This figure also includes the amortization of costs associated with drugs that fail to be approved.

Obviously, the cost makes it difficult for a small nanotech company to develop new drugs without substantial partnerships with established pharmaceutical companies. There are, nonetheless, a few nanotech drugs based on buckyballs or dendrimers that are under development; only one, from Starpharma, has entered clinical trials. Nanotech is expected to make its first impact on the drug delivery. Indeed, nanotech drug delivery techniques are already employed in some approved drugs.

Delivering Drugs

Drugs can be swallowed, injected, inhaled, or adsorbed through the skin. But having penetrated into the body, not all drugs are efficiently distributed to where they need to go (Table 12). A major problem is solubility. Oil and water, as we know, don't mix. Likewise, some chemical compounds are soluble in water, whereas others are happier in oil or oily substances, such as fat. Compounds that are sparingly soluble in water are difficult to deliver in pill form.

Table 12 Drug-delivery properties that can be obtained by nanoencapsulating drugs.

Timed release: the drug is delivered slowly over a period of time.
Rapid release: the nanoparticle improves solubility
pH-dependent release: the nanocapsule breaks down at a certain acid content, such as that found in cellular lysosomes
Ultrasound release: the nanocapsule or liposome is dissolved by a certain ultrasound frequency, releasing the compound or protein.
Magnetic release: capsules or thin layer releases drug compounds in response to a magnetic field
Target-mediated release: the capsule releases drug after binding to a specific molecular target
Targeted delivery: the capsule binds to the surface of specific cells or tissues
Brain or nervous system targeting: the nanocapsule mediates transfer through the blood–brain barrier
Imaging: drug-containing particles may be imaged by MRI or other means

Drugs are often put into capsules in the form of tiny granules. Having more surface area, these granules dissolve faster in the stomach than would a whole pill. An extension of this idea is to make drugs in a crystalline form, but with nanoscale grains that are small enough to be taken up into the bloodstream. In a nanoscale crystal, every molecule is close to the surface, and so this method can be used even for compounds that dissolve in water very slowly. Elan's NanoCrystal method yields particles of about 100 nm in diameter. This is not exactly revolutionary technology; it is more a matter of grinding drugs a little finer than previously. Nonetheless, it has proved useful for the delivery of a couple of approved drugs – Rapamune, an immunosuppressant used for recipients of organ transplants, and Emend, a drug used to prevent nausea in chemotherapy patients.

The precious metal silver kills bacteria; this little sliver of information has been common knowledge in the medical profession since the ancient Greeks. Bulk silver, however, does not dissolve well, so its efficacy as an antimicrobial has been marginal at best. Thus, it has largely been displaced by modern antibiotics. Nucryst Pharmaceuticals, however, has now introduced a nanocrystalline form of silver that has re-energized this ancient medicine.

Antibiotics are cleverly designed molecules that interfere with a microbe's biochemistry, usually at a single site. Silver, by contrast, is a blunt instrument that interferes in several different ways with a bacterium's life style. As a result, it is difficult for a bacterium to develop resistance. Silver is able to kill vancomycin- and methicillin-resistant pathogens, which have spilled out from hospitals into the community in recent years to become a major problem. Nucryst's president Scott Gillis admits that is possible to "evolve" silver-resistant bacteria by cultivating them over generations in increasing concentrations of silver. So far, however, they have not been found in the wild.

Nucryst Pharmaceutical's nanocrystalline silver is a textbook example of the difference in behavior between nanophase material and bulk material. The antimicrobial action of highly disordered nanocrystalline silver occurs more rapidly (in as little as 30 minutes) and lasts longer than bulk silver particles.

Nucryst's first product was Acticoat, a dressing for serious burns that is impregnated with nanocrystalline silver. The product replaces a generic silver-containing cream that was only active for a few hours, after which it had to be scraped off (at considerable cost in agony to the burn patient) and reapplied. Acticoat, on the other hand, is good for a week and can be lifted off in one piece. Acticoat is now also used for serious wounds, including diabetic ulcers. These wounds, which usually form on the feet or limbs, can require amputation if the wounds do not heal.

Acticoat was named one of the top ten nanotech products for 2004 (even though it was introduced in 1998) by the Forbes Wolf/Nanotech report. Though manufactured entirely by Nucryst, Acticoat is marketed by health science giant Smith and Nephew.

The success of Acticoat in healing recalcitrant wounds led to researchers to speculate that an antimicrobial action was not necessarily the whole story. Sure enough, it turns out that nanocrystalline silver also has an anti-inflammatory action as – specifically, it suppresses the activity of two inflammatory cytokines, interleukin 12B and tumor necrosis factor alpha. Thus, Nucryst is exploring its use in inflammatory conditions, including atopic dermatitis and certain respiratory conditions.

NovaVax, another small drug company, has a different way of delivering drugs using nanotech. The company has created what it calls Micellar Nanoparticles (MNPs), submicron-sized particles that be suspended in water. MNPs are derived from so-called "amphiphilic" molecules, which are water-loving at one end, and mix with oil at the other end.

NovaVax scientists have shown that MNPs are able to incorporate pesticides, proteins, whole viruses, flavors, fragrances and colors. MNPs also have the ability to entrap alcohol-soluble drugs, such as steroids, and to deliver these drugs through the skin.

NovaVax's first target indication with MNP technology is the hormone replacement therapy market. It's first product, called Estrasorb (estradiol topical emulsion), is an estrogen replacement therapy for women that the company has recently brought to market. A second product, Andrasorb, a testosterone replacement therapy for women, is still in the testing phase.

American Pharmaceutical Partners is basically a very successful generic drug maker that concentrates on injectable drugs. However, its compound Abraxane is a generic drug with a difference. The nanoparticle-based delivery technology actually makes the drug, paclitaxel (the active ingredient of Taxol) more effective and less toxic for the patient. Paclitaxel and its derivatives (called taxanes) are the most widely used drugs for the chemotherapy of cancer patients. Unfortunately, paclitaxel is not soluble in water, which is what the body is mostly composed of. Thus, Taxol consists of paclitaxel mixed with an awful solvent called Cremaphor and alcohol. Patients must be dosed with steroids in order to tolerate the side effects of Cremaphor, which is even more toxic than paclitaxel itself.

Abraxane consists of paclitaxel enclosed in a nanoparticle built out of serum albumin, which is the most common protein found in human serum The albumin nanoparticle in Abraxane is about 130 nm in diameter, somewhat larger than the 50- nm pores in capillaries that allow the egress of small particles into tissue. However, tumors are known to have particularly leaky blood vessels, so this larger size may actually promote the concentration of Abraxane specifically within tumors.

Because Abraxane does not use Cremaphor, patients do not have to be pretreated with steroids, and higher concentrations of paclitaxel can be given to patients. This higher concentration appears to make the drug more effective.

A natural transport pathway exists in the endothelial cells lining blood vessels whereby cells bind albumin to receive small molecules that bind loosely to the protein.

"Targeting endothelial cell transport is a novel approach to increasing drug concentration in the tumor and provides the opportunity to exploit a natural biological pathway the malignant cell inherently uses to supply itself nutrients and energy for rapid growth," says Patrick Soon-Shiong, President and CEO of American Pharmaceutical Partners. "Using this mechanism, the target would be shifted from the tumor cells themselves to the specific albumin receptor on the blood vessel wall of the tumor neovasculature. Abraxane represents the first example of the albumin-bound nanoparticle technology platform which may have the potential to exploit this natural biological pathway."

In early 2004, Abraxane was approved for the treatment of metastatic breast cancer and it is likely to be used in the future for a wide variety of tumors that respond to taxol.

Skye Pharma is a drug delivery company based in London that is developing many modes of delivery for drugs: oral, inhalation, or through the skin. Skye partners with many pharmaceutical companies that uses its delivery technologies. For instance, it receives royalties on GlaxoSmithKline's well-known antidepressant, Paxil. Skye enhances the solubility of certain drugs by incorporating the drug into very small particles. One variety is a solid lipid (a kind of fat) particle, which can be 40 to 1000 nm in diameter, which can be used to deliver drugs that are insoluble in water. Another drug delivery system uses nanoparticles composed of polymers. For more water-soluble drugs, Skye uses "Dissococubes" – a stabilized colloidal suspension of drug with standard-sized particles in the submicron range. The small size and greater surface area allows these particles to dissolve more easily.

Flamel Technologies is a French company named after a famous alchemist, Nicholas Flamel, who is supposed to have discovered the Philosopher's Stone, having read and apprehended a manuscript written by a man called Abraham the Jew, apparently the Biblical patriarch. Flamel Technologies does not transmute lead into gold; instead, they try to profit from various drug delivery technologies. Flamel, which apparently has a weakness for ancient mythology, has developed a technology called Medusa for the delivery of native protein drugs. Medusa, if you will recall, was a beautiful maiden who had the audacity to compare her physical attributes to that of the goddess Athena. Athena got royally ticked-off and turned Medusa's golden tresses into so many hissing snakes.

Flamel's Medusa is a nanoparticle composed of snake-like polymers of the natural amino acids leucine and glutamate. Leucine is oily and resides in the center of the particle, whereas glutamate loves water and winds around the outer surface. The combined particle has the ability to capture and stabilize many proteins and peptides (very small proteins, like insulin). Once injected, Medusa particles allow the slow release of peptides or proteins into the body.

Flamel has begun clinical testing of the Medusa particles ability to deliver a form of insulin and also interferon, a critical part of the immune system's viral defense.

Many entrepreneurial companies are trying to extend to nanoparticle drug delivery concept in new directions. For example, Advectus Life Sciences has developed what it calls Nanocure, a polymeric nanocapsule that has the unique ability to pass through the blood–brain barrier. In this way it can deliver high concentrations of drugs to the brain, without the need to drill through the cranium. Kereos is developing a platform based on a "targeted nanoparticle" in which a lipid "membrane" encloses a core made out of perfluorocarbon. Antibodies or recognition peptides incorporated in the surface allow the particle to be targeted towards cell-surface molecules characteristic of tumors or particular tissues. These nanoparticles could carry either an MRI contrast agent that allows the visualization of tissues by MRI scanning, or they may contain drugs. Thus, the nanoparticle can serve as means both to image cardiovascular abnormalities or tumors and to deliver drugs specifically to these pathological areas.

The National Cancer Institute (NCI) has big dreams. Its goal is "to eliminate suffering and death from cancer by 2015." To achieve this (unlikely) goal, the Institute would like to find ways to treat cancer that are much smarter and less toxic than current chemotherapy agents. These agents would be multifunctional.

One module would target the device particularly to tumor cells. Another module would contain a contrast agent that would make it visible by MRI imaging, X-rays, or would glow with fluorescent light in response to laser stimulation. Another module would contain a cell-killing drug that could be released by light activation or possibly by magnetic fields. This is not just a dream, but also a field of investigation. In particular, NCI is funding James Baker, from the University of Michigan, to research these possibilities. In Baker's view, these modules would be composed largely of dendrimers, branching polymers that assemble themselves into globular sub-units similar to proteins. These multi-functional devices could not accommodate a miniaturized being, like Ms. Welch, but they would have some of the attributes of miniature machines.

"Twenty years ago, without ...crude chemotherapy I would already be dead," commented Nobel Prizewinner and nanotechnologist Richard Smalley before a congressional subcommittee. "But twenty years from now, I am confident we will no longer have to use this blunt tool. By then, nanotechnology will have given us specially engineered drugs... that specifically [target] just the mutant cancer cells in the human body, and [leave] everything else blissfully alone.... I may not live to see it. But, with your help, I am confident it will happen."

The Fantastic Voyage is still ahead of us.

Medical Imaging: X-Ray Tubes

A promising technological application of carbon nanotubes is for field emission devices, a property that has been exploited by Otto Zhou of the University of North Carolina in the creation of an X-ray device.

There is a dramatic enhancement of the electric field at the end of a nanotube, owing to its small diameter and sharp geometry. As a result, electrons can be extracted from the nanotube tip by low voltages applied between the tube and an electrode. Electron field emission sources based on carbon nanotubes have several advantages over the silicon or metal tips currently used: low turn-on voltage and high current densities; high chemical stability against degradation, thus non-stringent requirement for the level of vacuum required vacuum level.

Experiments by the North Carolina team have shown carbon nanotubes can generate intense electron beams that can be used to bombard a metal "target" to produce X-rays. Researchers say they have demonstrated that their cold-cathode device can generate sufficient X-ray flux to create images of extremities such as the human hand. One advantage of using carbon nanotubes is that machines incorporating them can work at room temperature rather than the 1500 or so degrees Celsius that conventional X-ray machines now require and produce.

"We already have taken pictures of human hands and fish that are as good as standard X-rays," Zhou says. "We think our images eventually will be clearer than conventional ones since we have a more pointed, tunable source of electrons. That would help doctors, for example, get more useful information from pictures of broken bones, for example."

Zhou is a founder of Applied Nanotechnologies, which is working with manufacturers to turn their discovery into working machines. Being able to miniaturize X-ray devices could have more major benefit, including allowing technicians to take X-rays before ever leaving the scenes of accidents. In addition, the new X-ray technology may allow manufacturing of large-scale X-ray scanning machines for industrial inspections, airport security screening and customs inspections.

Nanoprietary Inc. has also developed a nanotube field emission device, a "cold cathode" used to produce X-rays. Oxford Instruments has used these in portable X-ray spectrophotometers, possibly the first commercial use of carbon nanotubes. Eventually, the companies expect to use cold cathodes for medical applications, particularly for miniature X-ray tubes. X-ray tubes as narrow as 4 mm are possible, according to Medi-Rad, an Israeli company. Nanoproprietary is currently supplying carbon nanotube cold cathodes to Medi-Rad and a Japanese company, although commercial products have yet to be developed.

This new X-ray tube will have applications in dentistry and may also improve brachytherapy treatments – localized radiation to kill tumors. These new radiation therapy devices should achieve overall lower cost and greater safety due to the absence of radioactive isotopes.

"Medical applications represent a potential for high volume production of carbon nanotube cold cathodes," says Zvi Yaniv, President and CEO of Nanoproprietary.

According to Oxford, because the cold cathode does not vaporize, the end-of-life mechanism as found in conventional X-ray tube technology does not exist. Standard X-ray tube technology typically will age and fail as the filament either fails or as a result of loss of internal vacuum associated with filament outgassing. Nanotube-based cold cathodes promise a much longer lived X-ray tube for use in portable environments.

Although nanotube-based cold cathodes have technical advantages for portable applications, they do not yet hold economic advantages because of the relatively high cost of nanotubes. However, these costs should come down as mass production of nanotubes increases. Tests that analyze tissue generally do so after the cells have been removed from the body.

A vital need is for devices that can examine the integrity of tissue within the body. To some degree, this has been solved by optical fiber endoscopes that can be threaded through blood vessels or other orifices. One company, Givens Imaging, has invented a microelectronic, disposable, capsule-size camera that can actually be swallowed and will generate pictures, which are radioed to a receiver as the camera follows its path through the digestive system. This is a microscale, not a nanoscale device. Obviously, a nano version that could traverse the bloodstream would have interesting applications, for example imaging atherosclerotic plaques. One academic researcher hopes to achieve a nanoscope so small that it can actually be inserted inside cells.

Making the World Safe for MRI (Plus some other Stuff)

Biophan is a small company with nanoscale technology and a big mission – to make MRI scans safe for all those who need them. As it happens, the company previously embarked on an even bigger mission – to develop a therapy for AIDS, although that idea has been put on hold. The story of Biophan is full of twists and turns, but its long winding road seems headed for paydirt. Which is only appropriate because, technically speaking, Biophan started out as Idaho Copper and Gold.

Here is the story: Mike Weiner, entrepreneur, part-time inventor, and full-time tech junkie, had formed a company called Technology Innovations (TI). A potential investor, Wilson Greatbatch, himself a multi-faceted inventor of the archetypal Thomas Edison variety, read the business plan for TI. Greatbatch is credited with inventing the first implantable pacemaker, and also developed the lithium iodide battery used in most implantable microelectronic devices. Not satisfied with just being an electronic engineer, Greatbatch had also delved into molecular biology. His explorations resulted in several patents related to the use of nucleic acid antisense molecules to treat HIV infection. Greatbatch thought that TI could help commercialize his patents. Mike thought it was worth a try.

Together, Mike and Greatbatch created a company called GreatBio Technologies. Biophan was later merged into an older company, a shell of its former self, called Idaho Copper and Gold. As the latter was already a public company, this "reverse merger" gave GreatBio tradable common stock. However, Greatbatch's original company, called imaginatively Wilson Greatbatch, Inc. objected to the name GreatBio as well as the stock symbol GBTI. In 2001, GreatBio became Biophan, as in Bio (technology), pha(rmaceuticals) and n(anotechnology).

Not long after Biophan came into being, Mike accompanied Wilson Greatbatch to a dinner put on by the National Inventor's Hall of Fame, into which Greatbatch had been inducted. There, the two had a discussion with Ray Damadian, inventor of the MRI scanner that has revolutionized medical imaging. Damadian asked Greatbatch when he was going to get around to creating a pacemaker that was safe to use in MRI scans.

MRI machines use very powerful magnets and electromagnetic radiation to produce their images. As you might remember from high-school science courses, magnetic fields induce electric current in conducting materials – for example, the leads and electrodes used in pacemaker and other microelectronic implants. A pacemaker can go haywire during an MRI examination, causing the heart to beat wildly and leading, in documented cases, to death of the patient. There is also a heating effect, which can damage tissue adjacent to a metal implant. This is a serious problem for patients not only with pacemakers, but also with defibrillators, neurostimulators, drug pumps, and certain prosthetics. As a result, MRI is effectively eliminated as a diagnostic tool for a significant percentage of the patient population most likely to need them – older folk who are the major consumer of medical services. Metal parts that many of us older cyborgs have accumulated – screws, plates, staples, and stents – may not disqualify us from MRI exams, but they do tend to mess up the images surrounding the devices. All of these mechanical parts for people add up to about a $10–12 billion annual market worldwide. Both MRI manufacturers and medical implant manufacturers would like to see the restriction of MRI use disappear.

MRI could also be used to guide catheters and guidewires, and endoscopes into the patient for minimally invasive procedures – if the devices did not heat up and burn the patient, not to mention the surgeon.

What could Biophan do to make MRI scans safer? The first idea that they came up with was a photonic fiber-optic system to replace the metal leads. This is feasible; however, it turns out that it would require more battery power. Also, it would entail getting a whole new technology through the FDA's labyrinth of regulations. Another idea presented itself through Mike Weiner's maze of connections.

Some people have catchphrases that run through their conversations. With President George Bush, who likes to keep things simple, the phrase is "*hard work*– as in "the war on terror is *hard work*", "being President is *hard work*", "debating John Kerry is *hard work*." With Mike Weiner, the phrase is "*Another company that I'm associated with* ... ." Besides being the CEO of Biophan, Mike sits on the boards of Biomed Solutions, LLC, Technology Innovations, LLC, Speech Compression Technologies, LP, Nanoset, LLC, and Nanocomp, LLC. Most of these firms are overlapping; Mike calls this collection of companies which share management and resources a *keiretsu*, a Japanese term applied to companies networked around banks and trading companies. In Mike's keiretsu, Technology Innovations begat Biomed Solutions, which then begat Biophan (which has now conceived TE-Bio out of an incestuous alliance with Biomed Solutions – more on that below). Nanoset, LLC owns many of the patents and patent applications licensed to Biophan. TI and Biomed Solutions both own significant equity positions in Biophan.

When Mike Weiner gives a keiretsu corporate Christmas party and the eggnog starts flowing, you can imagine that all sorts of weird and wonderful techy conversations get started. Wilson Greatbatch started talking to Xingwu Wang, an Alfred University professor associated with Nanoset about the safety problems that electrical leads cause during MRI procedures. Xingwu says, "Well, I think I can fix that."

Wang's idea was to coat the leads with a thin film of magnetic nanoparticles. These essentially act as a shield, reflecting radio-frequency (RF) waves from the MRI machine and minimizing induced current. Another innovation was a high-pass RF filter that prevents high-frequency pulses from being passed on to the tissue.

While working with Wang's magnetic nanoparticles, it was realized that coatings could be "tuned" to give off very high or very low MRI "signatures." Therefore, it was possible to make pacemakers or other implants visible to the physician through MRI. These could also be used to make guidewires that are used to implant stents and other devices visible to the physician. Finally, the magnetic nanoparticles, by themselves, could be used as MRI contrast agents (Fig. 42). These agents make the contrast between tissue types greater in the MRI image, making things easier for the physician who has to interpret them.

Mike and his team analyzed the market for MRI contrast agents and found that 25 % of all MRI procedures use them (your author has had two such procedures, and both used contrast agents). The market for contrast agents is $800 million and growing, as MRI supplants other diagnostic technologies. Currently used contrast agents for MRI are based on gadolinium, a highly toxic substance, making such MRI procedures not without risk for the patient.

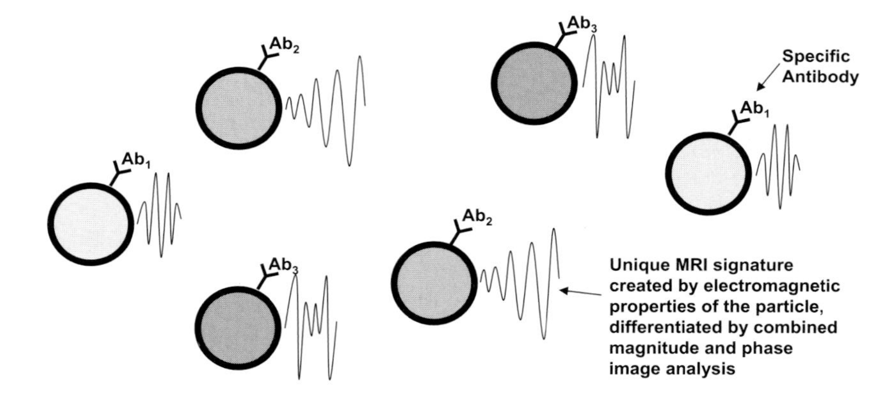

Each population of nanomagnetic contrast particles has a unique MRI signature (combined magnitude and phase information) and has a unique recognition molecule (Ab_1, Ab_2, Ab_3). Thus tissue targeting and image enhancement may be customized for specific disease diagnosis.

Figure 42 Nanomagnetic MRI contrast particles. Courtesy of Mike Weiner, CEO, Biophan, Inc.

Biophan hopes to be able to make their contrast agents tissue-specific. This could be done, for instance, by coating the particles with antibodies or receptor ligands that are bound only by certain cells. Because particles can be adjusted to give different MRI characteristics, it is conceivably possible to do multiplexing scans where different tissue types are labeled with different particles that could be distinguished in the scan.

At Nanoset, nanomagnetic particles have developed yet another talent – drug delivery. The total world market for drug delivery solutions is now estimated at around $40 billion. Drugs can be bound to targeted nanomagnetic particles that respond to externally applied magnetic fields. The magnetic field can be used both to aggregate the particles in one area of the body, and then can be modulated to cause the particles to release the drugs. If the particles are targeted to specific tissues, the position of the particles can be verified by MRI.

A similar technology may be used to reload drugs on drug-eluting surfaces; for instance, drug-eluting stents. Stents are the little wire cage devices that are inserted to keep arteries open to the heart. They are loaded with drugs to prevent "restenosis" – reclosure of the artery. Often, the plaque that restricts arteries will actually build up within the stent itself. With current technology, drugs are released passively from the stent coating until the drug is used up. The nanomagnetic particles may be used to recharge the stent coating with a new dose of drug. Alternatively, the magnetic nanoparticles could be incorporated into the stent coating. The drug within the nanoparticles could be released as needed by activation with magnetic fields.

Biophan now also has technology that allows plaque to be visualized within a stent as the result of its acquisition of Amris GmbH, a German company that had been a competitor in the field of MRI safe and image-compatible medical products. Biophan also has licensed a second means of visualizing plaque within stents from German MRI researchers Arno Bucker and Alexander Reubben. This is an application of "The Way According to Wiener." Acquire intellectual property regarding all potential solutions to a problem. This limits competition downstream.

Biophan has set up a new division called Nanolution to work on drug delivery systems. The president of the new division is John Lanzafame, a drug delivery veteran. He was previously president of STS Biopolymers, which was acquired by Angiotech Pharmaceuticals, the company that provides the drug-eluting coatings now used on stents sold by Boston Scientific. In fact, Biophan now has collaborative projects with Boston Scientific in progress.

"Another company that I'm associated with," says Mike, again, "had developed a thin film of thermoelectric material that could be used to drip charge batteries for medical implants." Since Biophan was already looking to enter the market for medical implants, this was a natural. Biophan and Nanomed Solutions formed a new company, TE Bio around this technology.

In a thermoelectric material, a current is induced in response to a gradient in temperature from one side of the material to the other. This thermoelectric effect has been known for a long time; in fact it was used to power the systems aboard the Voyager spacecraft, which continued to broadcast to Earth until it finally left the solar system, with its tiny time capsule addressed to interstellar travelers. However, nanoscale thin films have now reduced the temperature gradient required for the thermoelectric effect from 30 °C to about 3 °C, a differential that can be found within the human body. TE hopes to reduce the necessary temperature further, down to 1 °C.

TE-Bio expects that its thermal electric batteries for implants that will last as long as thirty years. Currently, when your pacemaker battery goes dead (after about ten years), you need a new pacemaker, which also requires another round of surgery and about $20 000 in medical expenses. But if you could keep the battery topped up using no more than body heat, you can get essentially a lifetime guarantee on your pacemaker. Besides pacemakers, the battery could be used for defibrillators, neurostimulators, and drug pumps. The market for defibrillators and pacemaker batteries alone is about $150 million.

In a masterpiece of technology outsourcing, Biophan has lined up NASA to do the development work on its biothermal battery. NASA, it seems, wants to have long-term health monitoring devices implanted into its astronauts. Biophan has licensed to NASA rights for use of the technology in space, while retaining earthbound commercial opportunities. Not a bad split from Biophan's perspective.

Biophan's not-quite-ready-for primetime slogan is "We don't make medical devices; we make them safe, and imageable for MRI. And now, we make other stuff too." Starting from a mission to make MRI procedures safer, Biophan has already diversified into contrast agents, drug delivery, and batteries. Biophan's intellectual property portfolio is burgeoning, with over 100 patents or patent applications owned or licensed.

Running a technology company, points out Mike Weiner, is not like traditional project management, because the time lines in development are not secure. Because one technology may run into an unforeseen roadblock, it is desirable to have a variety of possible solutions, and for that matter, a variety of potential applications. A technology company is not unlike a wildcat oil exploration, in Mike's view. There is always the risk of a dry hole. For this reason, one wants to own a lot of different properties.

Not long ago, at a Nanotech conference, I was having lunch with Mike and several of his associates. Mike was filling me on the highlights of his career. He was once a commercial fisherman and actually wrote a book on how to get a commercial fishing loan. From there, somehow, he got involved with word processing software, creating the award-winning Word-Finder thesaurus for PC and MacIntosh. He also did a stint at Xerox Parc, where he shepherded the spin-off of Microlytics, a software company. This experience came in handy when starting Biophan and his collection of interlocking companies, including most recently NaturalNano, a materials company (see Chapter 6). He was about to tell me about some new venture involving stem cell therapies when he was interrupted by a cell phone call.

"From commercial fishing to software to Biophan to stem cells? How does he do it?" I wondered aloud.

Sitting next to me at the table was Alexander Weis, Chairman and CEO of biopharmaceuticals company OncoVista and a long time Wiener associate. "We are limited by knowledge," says Weis, "Mike doesn't know what is impossible so he just does it."

Of course, Weis hasn't done so badly himself. The last company he co-founded, Ilex Oncology, was eventually bought by Genzyme in a transaction worth about $1.4 billion. I should be so limited.

Other Contrast Agents

There are many companies besides Biophan attempting to supply the $800 million plus market for MRI contrast agents, using several different nanoparticle types (Table 13). Contrast agents provide a clear picture during MRI procedures. Patients are injected with these agents to help physicians diagnose problems or diseases, especially of the brain or spine. The chief requirement of an MRI contrast agent is a lot of unpaired electrons that interact with the protons in water. Chelated gadolinium compounds are the leading MRI contrast agents; for example, a commercial compound called Magnevist works well because of its seven unpaired electrons. Magnevist magnetically attracts protons of water present in tissues, accelerating their relaxation between RF pulses. A faster relaxation leads to a higher signal intensity and therefore greater contrast in the MRI images.

Table 13 Nanoparticles as MRI contrast agents.

Type of agent	Company
Fullerene endohedral	Luna Nanomaterials
Perfluorate nanomicelle	Kereos
Magnetic nanoparticle	Biophan
Nanoparticle	Nanomed Pharmaceuticals
Dendrimer	Dendritic Nanotechnologies

Although organic chelates of gadolinium do a fair job of reducing toxicity, encapsulating the metal inside a fullerene might be even safer. Such an "endohedral" could offer additional advantages. For example, the trimetallic-nitride-containing endohedral C82-fullerenes, first reported by chemistry professor Harry C. Dorn's group at Virginia Polytechnic Institute & State University, Blacksburg, can accommodate three metal atoms inside each cage, potentially offering a more potent agent. These compounds are being commercialized by Luna Innovations, a Blacksburg VA research company. Luna's first focus is to develop MRI contrast agents that provide enhanced image resolution.

Other nanoparticles being considered for use as contrast agents include dendrimers (Dendritic Nanotechnologies) and perfluorate nanomicelles (Kereos) and other nanoparticles (Nanomed Pharmaceuticals),

Nanoshells for Therapy

Nanoshells are a class of nanoparticles made of extremely small gold-plated glass beads, approximately 100 nm in diameter, invented by Naomi Halas, a professor at Rice University. Nanoshells are able to absorb light at almost any wavelength, especially in the near-infrared region, which can pass through human tissue. By varying the thickness of the constituent metal layers on the surface of the nanoparticles, the color and light- absorbing properties of nanoshells can be accurately controlled. Making use of their light-absorbing properties, nanoshells have been tested for cancer therapy, drug delivery for chemotherapy and pain management within the body.

However, laboratory experiments have demonstrated that nanoshells can be targeted towards specific diseased cells, while leaving healthy tissues unimpaired. This can be done by injecting nanoshells tuned to absorb infrared light. Localization of the effect can be achieved by controlling both the site of injection and by targeting the infrared source to particular parts of the body, say a solid tumor. The heat generated in the nanoshells by the infrared destroys the tumor, without damaging surrounding tissues. This application is being commercialized by Nanospectra Biosciences of Houston, Texas.

Another potential application of nanoshells is to deliver drugs within the body, when and where it is needed. This can be done by attaching nanoshells to temperature-sensitive polymers that change shape when they are heated. These polymers have a critical solution temperature that is slightly higher than body temperature. When a strong outside infrared source is applied, the nanoshells heat up the hydrogel. Once the temperature exceeds the critical solution temperature, the hydrogel liquefies and the medication within is released.

Pumps

A California-based company, IMEDD (for Intelligent MicroEngineered Drug Delivery) is focused on tiny drug delivery pumps. The company has been working with Terry Conlisk, an engineer at Ohio State University, who has developed a computer model to help tiny implanted drug reservoirs to pump out drugs on demand. Pumping through nanopore channels is a technical challenge. Conlisk has developed a comprehensive computer model to address electrically driven fluids in channels.

If a fluid is positively or negatively charged and there is a like charge to the inner surfaces of a channel, the charges will repel each other. The result is that the fluid will flow down the channel.

In IMEDD's experiments, engineers were able to flush almost 0.5 nL of saline per minute through a channel only 7 nm wide, similar to computer predictions. In a 20-nm channel, the flow rate was almost 0.8 nL per minute. Although the computer models were based on electrical potentials as large as 6 V, the experiments showed that much smaller voltages could be just as effective. In practice, the voltage needed would depend on the size of the implantable device and the amount of drug that had to be dispensed.

By manipulating colloidal microspheres within customized channels, Alex Terray, John Oakey, and David W. M. Marr, from the Chemical Engineering Department at the Colorado School of Mines, have created micrometer-scale fluid pumps and particulate valves. They have described, in a *Science* article, two positive-displacement designs, a gear and a peristaltic pump, both of which are about the size of a human red blood cell. Two colloidal valve designs are also demonstrated, one actuated and one passive, for the direction of cells or small particles. The use of colloids as both valves and pumps may provide a link between fluid manipulation at the macro- and nanoscale.

Medical researchers would like to use nano-scale tubes to push very tiny amounts of drugs dissolved in water to exactly where they are needed in the human body. The roadblock to putting this theory into practical use has been the challenge of downsizing mechanical pumps sufficiently to do the job. In addition to the engineering challenge of building a nano-scale pump, there is the added complication of clogging by any biological molecule that can occur in valves small enough to fit a channel the size of bacteria. The solution – discovered by researchers at Arizona State University – is to create a system that does not rely on mechanical parts.

The ASU team of scientists and engineers reported in the American Chemical Society journal *Langmuir* in 2002 on a technique they developed to pull water up a tube tinier than a straw by shining a beam of light on the surface of the tube. This technological advance, referred to as photocapillarity, may one day find a use in nanotechnology applications, such as the targeted distribution of medicine in the body.

Bioengineering professor Antonio Garcia, and colleagues Devens Gust and Mark Hayes, professors in the ASU Department of Chemistry & Biochemistry, combined their bioengineering and chemistry skills to build upon the research on light-responsive molecules. With proceeds from a National Science Foundation grant, the researchers found a way of attaching the molecules to the surface and structuring the surrounding surface to control the spread of water.

"When we shine light just beyond the visible range, the light-responsive molecules attract water and trigger the advancement of water through the channel," says Garcia.

The Strange Case Of Nanobacteria

First, let's be straight. Nanobacteria are not really bacteria. About 7000 species of bacteria have been reasonably well described, and many times that number are presumed to exist, but all of have a diameter of a least 1 μm. Nanobacteria may not even be alive, in the sense that we understand the word. But they appear to replicate themselves – and they may be making us sick.

Nanobacteria were discovered in 1988 because a Finnish scientist, Olavi Kajander, was having trouble with tissue culture in his laboratory. The mammalian cells that he was trying to grow kept dying for no apparent reason. Determined to identify the problem, Kajander looked at some of his cultures under an electron microscope. Inside the cells, he saw some funny-looking things – some particles up to 100 times smaller than a bacterium, from 20 to 200 nm in diameter. Kajander called these novel organisms "nanobacteria," – perhaps an unfortunate choice of words, because microbiologists immediately dismissed the diameter as being too small to have the complex metabolism of a bacterium. Nanobacteria were dismissed as artifacts or, more charitably, as some type of new virus.

After ten years of criticism, Kajander tried to prove that nanobacteria contained ribosomal RNA, part of the protein-making machinery, and a prerequisite for life as we know it. Two years later, an NIH study proved that Kajander's RNA was actually a contaminant from a common laboratory bacterium. In fact, it was not clear at that time whether nanobacteria had any nucleic acids at all. Things began to look bleak for the nanobacteria hypothesis.

Nanobacteria might have been written off as an artifact but for a link between nanobacteria and human disease. Kajander found that nanobacteria had a hard outer shell composed of calcium phosphate. He began to wonder if his new bug was associated with kidney stones, which result from an aggregation of calcium compounds. With colleague Neva Çiftçioglu (don't you love Finnish names?), he found that some kidney stones have nanobacteria at their core. Moreover, direct injection of nanobacteria into rat kidneys resulted in the formation of kidney stones within a month. Austrian researchers also found nanobacteria within calcified deposits in ovarian cancer. The novel organisms have additionally been found associated with Alzheimer's disease and prostatitis, a calcification of the prostate gland.

A team from the Mayo Clinic also found that nanobacteria-like structures could be isolated in human calcified arteries and cardiac valves. Moreover, these particles could be stained with a DNA-specific dye and would incorporate radiolabeled uridine, a precursor of RNA. Not only that, but the nanobacteria could be cultured *in vitro* and appeared to replicate, albeit slowly. Nanobacteria double about once every three days, a glacial pace compared to garden-variety bacteria, which can double every 20 minutes given optimal conditions.

Despite the Mayo Clinic data, no one has yet succeeded in isolating and characterizing DNA or RNA from nanobacterium, a puzzling negative finding. It must be admitted, however, that relatively few researchers are working on the problem. Nanobacteria still suffer from an image problem, partly because of Kajander's overly aggressive efforts to promote them as a new form of life. They also lack a constituency. Microbiologists don't consider them to be bacteria; virologists are sure they aren't viruses. Kajander has come up with a new name for its strange bugs – "calcifying nanoparticles." Perhaps nanotechnologists will adopt them.

Kajander and Çiftçioglu aren't waiting. They have established a company, NanoBac Life Sciences, which calls itself the "world leader in the research of degenerative diseases stemming from nanobacterial infections." This is an aggressive claim since the role of nanobacteria in causing anything is far from proven. The company even markets Nanobac supplements, which are a form of chelation therapy. Its active ingredient is EDTA, which soaks up excess calcium and other divalent cations (doubly positive charged ions). Though EDTA may be effective in treating certain conditions caused by calcification, this does not necessarily imply that nanobacteria are the root cause of the condition.

Nanobac Life Sciences has developed blood tests to diagnose nanobacterial infection or exposure, and is also working to develop a pharmaceutical to treat nanobacterial infection.

Nanobacteria are found not only in human tissue but also, strangely enough, in geological formations. For instance, Everett Gibson of NASA's Johnson Space Center in Houston, Texas, and Colin Pillinger of Britain's Open University have found "nanobacteria" living 2 km under the Earth's surface in the basalt rock of the Columbia River Valley. Therefore, though they may be human pathogens, nanobacteria are not necessarily obligatory parasites because they can apparently replicate inside of rocks. It should be pointed out, though, that all nanobacteria may not be the same. The comparisons are based on size and morphology alone, as there are no genetic data.

The situation became even stranger as a result of a study undertaken by Çiftçioglu and David McKay, from NASA. The space agency knew that astronauts on extended missions had an enhanced tendency to develop kidney stones. They wanted to know whether nanobacteria, the presumed seed for kidney stones, grew faster under zero-gravity conditions. Sure enough, they did, from 3.2 to 4.6 times faster, depending on culture conditions. NASA and Nanobac are now working on methods to characterize kidney stone formation with some the same tools used to study moon-rocks.

Perhaps then the reason that nanobacteria grow so slowly on Earth is that they don't really belong here. Claims have been made of fossilized nanobacteria within meteorites. Has the Earth been seeded with an alien life form – nanobacteria from outer space? Swarms of rock-eating, cancer-causing bugs invading the galaxy – sounds like the next Michael Crichton novel. Stay tuned.

Medical Diagnostics

The different ways that people can be prodded and poked to see what, if anything, is wrong with them increases every year. Table 14 provides a by no means exhaustive list, based on the level of resolution. Nanotech/biotech analysis can supply some answers to questions posed in most of these categories.

Table 14 Medical analysis at different levels of resolution.

Population	Individual	Organs	Tissue	Molecular
Epidemiology	Vital signs	X-ray	Histochemistry	Gene sequencing
Genetic traits	Medical history	Ultrasound	Blood cell counts	Protein sequencing
Endemic diseases	Family history	MRI	PAP smears	ELISA
Historical statistics	Symptoms	CAT scan	Biopsies	Expression microarrays
	Allergies	Pulmonary tests		Cell markers
	Blood chemistry	Heart catheterization		Affinity microarray
	Urine chemistry	Stress test		Mass spectrometry
	Saliva	ECG (EKG)		Enzymology
	Stool	Endoscopy		Pathogen identification

Certain genetic traits are known to be present in ethnic populations: Tay–Sachs in the Ashkenazi Jews of Eastern European, or sickle cell anemia in blacks. These traits can be confirmed by tests for single nucleotide polymorphism, which have now been automated to the point of a 30-minute procedure.

Diagnostic tests, as discussed in the section below, are being developed that will simplify and vastly increase our ability to analyze solutes, protein, nucleic acids, and other chemicals in blood, urine and the environment and thereby gain better diagnostic tests for disease.

Smart Dust

In a Defense Advanced Research Projects Agency (DARPA)-supported project, researchers at the University of California, San Diego have developed dust-sized chips of silicon that allow them to rapidly and remotely detect a variety of biological and chemical agents. Their work was reported in the October 1, 2002 issue of the journal *Nature Materials*.

"The idea is that you can have something that's as small as a piece of dust with some intelligence built into it so that it could be inconspicuously stuck to paint on a wall or to the side of a truck or dispersed into cloud of gas to detect toxic chemicals or biological materials," says Michael J. Sailor, a professor of chemistry and biochemistry at UCSD who headed the research effort. "When the dust recognizes what kinds of chemicals or biological agents are present, that information can be read like a series of bar codes by a laser that's similar to a grocery store scanner to tell us if the cloud that's coming toward us is filled with anthrax bacteria or if the tank of drinking water into which we've sprinkled the smart dust is toxic." The "bar code" on the silicon dust particles is basically a specific wavelength of light, or color, reflected from their surfaces after thin films layered on the silicon chip chemically react to a specific chemical or biological agent.

The scientists start with silicon wafers similar to those used in the manufacture of computer chips, and then "encode" them by generating layers of nanometer-thick porous films on the wafers using a special electrochemical etch. This layered structure on the dust-sized particles, which are created by breaking apart the wafer using ultrasound, imparts unusual optical properties to the particles. Referred to as photonic crystals, these micron-sized particles are able to reflect light of very precise colors, each one of which can be thought of as a single bar of a grocery store bar code. "When you're looking for chemical or biological warfare agents, you're going to want to search for thousands of different chemicals," says Sailor. "Since the particles can be encoded for millions of possible reactions, it's possible to test for the presence of thousands of chemicals at the same time."

The encoding that takes place in these particles provides monochromatic colors that are so sharp from the visible to the infrared that a laser can read thousands of distinct colors corresponding to separate chemicals. In this way, the UCSD researchers say these coded particles can perform thousands of biochemical assays in a small beaker or a Petri dish, which should be useful in many medical and research applications, such as the discovery of new drugs, the diagnosis of disease and the controlled release of therapeutic drugs.

Because the smart-dust chips are fabricated from silicon, they can be made using existing computer chip technology. The compatibility of porous silicon with living cells and the long-term stability and non-toxicity of this material may make them especially useful in biomedical applications.

For example, if a patient is feeling ill, his or her blood sample could be sent to the laboratory for screening. DNA probes for various types of infectious diseases could be coded with the crystals, and these probes could be mixed in with the patient's blood sample. If the blood sample binds with one of the probes, its crystal code will exhibit a pattern that identifies the probe, and thus diagnose the disease that the patient carries.

For the detection of chemical and biological warfare agents, the advantages of smart dust are numerous. Not only are the smart-dust crystals small in size, inconspicuous and capable of detecting thousands of possible agents at once, but they can detect potentially hazardous compounds remotely from a distance. Unlike grocery store scanners, which typically must read bar codes only inches away, Sailor and his group have been able to get their laser to detect the color changes in the smart dust 20 meters away, the length of the hallway outside their research laboratory. The group hopes to extend this beyond 0.8 km.

NanoArrays for Biomedical Assays

Bioforce Nanosciences has introduced NanoArrays, a nano version of microarrays commonly used for gene expression assays purposes. These arrays consist of particular gene sequences spaced at intervals in a grid along a glass slide. NanoArrays, which are prepared using dip-pen lithography (see Chapter 3), use approximately 1/10 000th of the surface area occupied by a conventional microarray, allowing greater sensitivity and less use of precious resources. Using a dedicated atomic force microscope, NanoReader as the readout tool for NanoArray, allows sensitivity to the single molecule level. Basically, the NanoReader picks up the increase in mass caused by hybridization to the target or formation of an antibody complex. Molecular interactions can be characterized in air or in liquids in minutes and with no molecular tags – no fluorescence, radioactivity or enzymatic reporters are necessary.

NanoArrays are suitable for the analysis of peptides, proteins, nucleic acids and small molecules. In practice, however, it is not clear that large proteins can be spotted using dip-pen lithography without significant denaturation.

NanoInk, an Illinois-based company, is also interested in developing nanoarray technology for drug discovery-type assays. NanoInk was founded by Northwester University professor Chad Mirkin, who invented dip-pen nanolithography.

Moving Water Around, a Little at a Time

A company called Picoliter in Mountain View, CA, developed a technology for moving and analyzing small amounts of liquid using ultrasound energy. The quantity that is transferred depends on the frequency of the sound wave used to transfer it. This technology provides the ability to dispense fluid samples with no contact and to reliably transfer volumes as tiny as a picoliter (pl = 10^{-12} L). For comparison, a dewdrop contains about 50 μL of water, or 50 million pL. So 1 pL is not very wet.

The Picoliter liquid handling technology may have broad applications in the life science tools market, including microarrays, dispensing equipment, and living cell transfer devices. Drug companies are interested in performing biochemical assays in as small a volume as possible because this results in lower volumes and therefore costs of expensive reagents used in the assays. Scaled-down assays also have an advantage in increased sensitivity and speed.

Picoliter successfully installed of its acoustic drop ejection systems at Hosokawa Micron Group in Osaka, Japan. Picoliter and Hosokawa are working to together to further develop applications of the ADE technology for the production of small, uniform and multi-layer particles for pharmaceutical manufacturing.

Labcyte, a company that manufactures robotic equipment for drug discovery assays, was so impressed with Picoliter's acoustic transfer technology that they bought the company in November 2003.

Nanoscale Antenna Controls DNA

Researchers at MIT can control the hybridization of DNA via a nanoscale radio antenna achieved the feat by chemically attaching a gold nanoparticle to a DNA molecule. The DNA was placed in a solution surrounded by a coil that generates an alternating magnetic field, which heats the gold antenna by induction. When the radio waves stop, the heat dissipates and the process is reversed. The DNA molecules dissipated the heat in less than 50 billionths of a second.

The researchers' method leaves molecules surrounding the targeted DNA molecule relatively unaffected. The 1.4-nm metal particle acts as a point heat source that heats anything in its vicinity.

"Induction heating is used in industry as a non-contact method of heating metals, and it can penetrate tissue," notes Kimberly Hamad-Schifferli of the MIT Media Lab.

The method could also be used to manipulate other types of protein molecules. It is relatively easy to attach a nanocrystal antenna to any type of protein, and the heat could be used to affect processes such as enzymatic activity, biomolecular assembly, and gene regulation.

Artificial Joints

As we grow older, the body parts that wear out first are generally our joints. The biggest culprit is osteoarthritis (OA), which is frequently blamed on ordinary "wear and tear," as if our joints had a certain mileage guarantee beyond which things just naturally break down. Though that may seem like common sense, on reflection, it doesn't hold water. "Young" joints hold up very well to repetitive motion. A professional baseball pitcher will stress his arm more in the few brief years of his career than a non-athlete will in the course of his entire life. Yet pitchers almost always retire long before arthritis claims their joints. Age, not usage, is a more general factor in the decay of joint function. OA begins in most people between the ages of 40 and 50 years, and is nearly universal beyond the age of 70 years. The problem is really that cartilage is not replaced in older people, probably because of a deficit of stem cells.

An even more serious condition is rheumatoid arthritis (RA). While osteoarthritis is a disease of the middle-aged and older, RA can strike even juveniles and lead to considerable disability.

I am indebted to rheumatoid arthritis sufferer Liz Whelan for this subjective opinion on the cause of her condition:

"Perhaps I'm presumptuous, but I believe I have figured out the true cause of the disease (RA). Gravity.

I first considered this theory while walking down the street on a fine spring day. Around me, happy New Yorkers were positively bouncing down the sidewalks, light as a feather, while I grimly trudged, one heavy step at a time. Now my scale tells me that I'm a mere 130 pounds and my eyes tell me that my fellow pedestrians are, by and large, well, larger. So how come they're practically air-borne while I can barely keep from stumbling to my knees? The only possible answer is that the specific gravity around my body is greater than theirs.

Once you have been infected with increased gravity, your joints are slowly crushed by the increased weight you bear. The depression you feel is actually the weight of your skull pressing against your brain.

I have further observed that the increased gravitational pull extends to the objects around me. I have only to walk by an object resting at the edge of the table at it seems to leap to the floor of its own volition. I prop my cane securely in the corner, and by the time I turn around, it is clattering on the floor. Three out of four objects I pick up immediately fall out of my hand. Not my friends, because of all the inflammation, swelling and pain, but because of the increased gravity!"

Though Liz has eloquently described what it feels like to have RA, physicians do not generally accept her increased gravity theory. RA is an autoimmune disease – the patient literally becomes allergic to him/herself, or at least to the cartilage surface lining his/her joints, the near friction-less bearings that make movement possible. The inflammation, swelling and pain are caused by the attack of the patient's lymphocytes and macrophages. Eventually, the cartilage breaks down, the bones rub against one another, and movements become painful, sometimes impossible.

The end-stage treatment for an arthritic joint, like cancer, is surgical removal. In some cases, like the small toe joints that are really dispensable, and vertebral joints, which have some redundancy, the bones are simply fused together. In most cases, the joint is replaced with a manufactured equivalent that is engineered to approximate the movement of the natural joint.

As arthritis strikes any number of joints, it is not uncommon for patients to have two, three, or even more implants. A typical case is RA patient Lori J., who reports, "I had my toes fused December, 1993, my left knee July 1995, my right shoulder June 1996, and my right knee in May 1998." She had turned just 41 about the time of her last joint replacements, and it is likely that she will eventually have to have several more joint implants in her lifetime.

Annually, in the United States, there are about 300 000 knee replacements and about 150 000 hip replacements. Shoulders, elbows, thumbs, and even big toe joints can also be replaced. Not only do they usually work as well as the one's that Nature designs, but they have the advantage of being devoid of nerve endings. Alas, there is a problem. Eventually, artificial joints stop working. The problem isn't usually with the titanium or ceramic materials that the joints are made of; the problem lies in at the interface between the bones and these foreign materials Eventually, the artificial joint becomes loose, leading to loss of function. Sometimes, an operation can fix the problem and sometimes not.

Because artificial joints ultimately fail, most surgeons prefer to reserve joint replacements for older patients. Some RA patients, however, have little choice, if they want to maintain at least minimal mobility: as a consequence many people with RA become permanently disabled before they achieve senior citizenship.

One large problem for the interface between artificial joints and bones is that the former are just too smooth. Just roughing up the artificial surface improves the binding of bone at the interface. Osteoblasts, the cells responsible for making new bones, like to crawl into the little crevices to remodel bone around the implant. To create the roughness at the surface, the joints are sometimes coated with calcium phosphate crystals, also known as hydroxyapatite. Currently, these crystals are in the micron range. Titanium powder is sometime used instead of calcium phosphate.

Titanium or other materials used in implants typically have surface features in the range of microns. However, bone is a nanophase material. The calcium in our bones is in the form of nanoscale calcium phosphate (also called hydroxyapatite) crystals organized by collagen proteins. Typically, the crystals have dimensions about of 50 nm in length and 5 nm in diameter. These are strung along collagen fibers that are about 300 nm long but only 0.5 nm in diameter. In order to maximize the surface area between bone and implant, it is necessary to make the implant nanophase as well, at least at the surface that is joined to bone.

Current joint coatings have an unfortunate tendency to delaminate, resulting in a loss of contact between joint and bone. One nanotech company, Inframat, hopes to improve the coating on artificial joints simply by replacing the current microscale coatings with nanoscale hydroxyapatite particles. It is easier for the FDA to approve this sort of incremental change, if it can be shown that the nanoscale coatings are indeed an improvement. Nevertheless, it is expected to be several years before such coated implants are on the market.

Researchers at Purdue University, led by Tom Webster, have shown in a series of experiments that bone cells in Petri dishes attach much better to materials that possess an architecture with smaller surface features than are found on the conventional materials used to make artificial joints. The smaller features also stimulate the growth of more new bone tissue, which is critical for the proper attachment and maintenance of artificial joints once they are implanted.

Webster has improved on his technique by creating surface features with carbon nanotubes. A single-walled nanotube is all surface, since every carbon atom within it is on the surface. It turns out that the nanotubes work best for bone cell attachment if the nanotubes are aligned. The aligned tubes resemble natural collagen fiber which, like carbon nanotubes, are long and very thin, and tend to align in one directions. In some experiments, Webster has aligned the nanotubes along the a titanium surface, a common material used in artificial joints. Osteoblasts adhere to the nanotubes almost immediately.

Unlike the point of contact with bone, the working surface of the artificial joint should be smooth to function with as little friction as possible while hard enough to resist wear. At the University of Alabama, Yogesh Vohra thinks he has found the perfect material: nanophase diamonds. This is another one of those ideas that struck a prepared mind after a lucky laboratory accident. His team was trying to grow diamond crystals in a gas reactor that held a combination of methane and hydrogen. One day, the reactor sprung a leak, allowing air into the chamber. The nitrogen in air prevented the diamonds from growing into large, faceted crystals. Instead, the diamonds made a smooth nanoscale coating over the surface of the metal substrate. Being diamond, such a surface is incredibly hard and resistant to scratching, and yet is smooth enough to minimize friction. Vohra has already tried coating metal mandibular joints (jaw implants), but the coatings have not yet been tried in human experiments.

Artificial Organs

Cochlear Implants

The brain–machine interface, a staple of cyberpunk science fiction, like William Gibson's *Neuromancer* [2], would allow the human brain to interact directly with computers and other machines. Although jacking a computer into your head sounds very far out, the fact is that such devices have existed for years in the form of cochlear implants, which provide a functional sense of hearing to people who would otherwise be essentially deaf. Cochlear implants bypass the middle-ear altogether, and feed auditory input, massaged by computer programs, directly to the auditory nerve.

About 50 000 people around the world are already cyborgs of a sort because their inner ear has been replaced with a cochlear implant. These devices have an increasingly good record of providing hearing to the otherwise profoundly deaf. They do so by means of electrodes that contact the cochlea at various locations. Nerve fibers are tuned to a different pitch depending on where on the cochlea they reside. High frequencies are picked up at the base of the cochlea, whereas lower pitches are picked up toward the center of the spiral-like organ.

A fairly small number of electric contacts is necessary in order to recognize speech, but obviously the larger the number, the greater the potential resolution. Modern devices now have as many as 125 contacts. Software also helps to decode the incoming sound wave.

Cochlear implants, as of now, are not nanoscale devices, although later devices may benefit from nanoscale engineering. Implants that must interact with many neurons, such as a retinal implant, have to be built with electrode arrays. In the case of the retinal implant, the larger the number of electrodes, the greater the visual acuity that is possible. Each electrode contributes something to the image similar to the pixels of a computer display, which may have a million pixels. Nanoscale engineering may eventually allow an ocular implant that essentially replaces the eye. Current electrode arrays must be increased by several orders of magnitude to accomplish that feat.

The current models of cochlear implants are sufficient for people to recognize voices, understand conversations in crowded rooms, and speak over the telephone. The one downfall so far has been in the appreciation of music, which tends to come across as harsh and grating to the implantee. It may be a while before a Beethoven symphony retains all of its majesty, or the sultry voice of Norah Jones its natural appeal, when transmitted through a cochlear implant.

A new generation of middle-ear hearing implants is under development by NanoBioMagnetics. The company is attempting to develop and commercialize magnetic nanoparticles as middle ear implants for hearing restoration. Phase 1 of an NIH SBIR grant will demonstrate feasibility of preparing hermetic ferromagnetic nanoparticles and attaching them to middle ear ossicles, the tiny bones that transfer a sound impulse from the eardrum to the inner ear. Through the use of an electromagnetic coil, the magnetic particles can be used to generate force through ossicular chain. This is the principle behind a new sort of hearing aid that NanoBioMagnetics hopes to develop. They have done proof-of-principle experiments using rats.

The auditory brainstem implant (ABI) was invented for people in whom a cochlear implant is not a viable option. It was developed by Cochlear Corp. in collaboration with the Huntington Medical Research Institute to provide some auditory function for people suffering from a rare, genetic disease, neurofibromatosis type 2. These unfortunates develop bilateral tumors affecting the auditory nerves. Removal of the tumor results in profound deafness, which is the superior alternative to death. About 40000 Americans suffer from this disease. Few have yet received an ABI implant, which is still an experimental device.

The ABI is implanted in the cochlear nucleus, which is part of the brainstem, not the inner ear. This nucleus is characterized by complex layers of neurons and axons organized to some degree by the pitch to which they respond. Unraveling that complex maze is what it will take before the ABI becomes as useful as a cochlear implant. Nevertheless, the National Institute of Nervous Diseases and Stroke (NINDS) has a program to do just that, primarily working with kittens.

Like cochlear implants, the current versions of the ABI are microdevices. The ABI, however, does not work nearly as well as the cochlear implant. At best, it augments the ability of some people to read lips and allows them to recognize some environmental sounds, like barking dogs or honking horns. It seems likely that a truly effective version will have to take advantage of nanofabrication to obtain the number of electrodes required before reasonable semblance of hearing can be duplicated.

The Artificial Retina

"Rotating through a 180-degree arc, the propulsion device collides inelastically with an orb approaching at 95 miles per hour, reversing its angular momentum and direction. Two image acquisition units monitor the electromagnetic waves reflected by the orb; these measurements are used by processing module to calculate the orb's new velocity and direction in real time. This information is seamlessly integrated with motor control of an orb-collection-device, otherwise known as a major league shortstop, allowing him to anticipate the bad hop, and prevent the ball from escaping into the outfield."

Our eyes, though they be the windows of the soul, have a certain mechanical aspect to them. They are very much like cameras. The eyelids are their lens caps. Though differently constructed, the lens of the eye functions to focus light very like the lens of a camera. The iris sets the F-stop, depending on light intensity. The light is focused on the retina, which is more like a digital recorder than like a chemical film.

In kind and quality, visual senses differ widely throughout the branches of the tree of life, from the compound eye of the honeybee, to the very acute eye of the raptor, to the nearly blind mole. Some starfish, as we have seen, have eyes. Even some bacteria can sense light and color.

Among mammals, primates and people, in particular, have the really premium quality vision due to one to two million rod and cone cells, the photoreceptors of our eyes, which allow for the incredible feats performed by our athletes. For the rest of us, as well as being a source of information, our rich visual experience accounts for a lot of our aesthetic appreciation of the world. Sadly, as we grow older, our sight dims noticeably. We need more illumination to see clearly, and the fine print gets fuzzier. An unfortunate minority descend into total darkness, due to trauma, glaucoma, congenital disease or just plain age.

Restoring sight to the blind has traditionally been the province of saints and holy men. A more pragmatic goal, for the present, is to restore a "useful visual sense" to the blind, sufficient for reading signs, perhaps, or walking down a crowded sidewalk. These tasks do not really require the full range of vision that most of us enjoy.

"Miracles of real-time visual behavior are performed by the common housefly whose brain is the size of a grain of rice," observes Christof Koch of the California Institute of Technology. A fly gets by on about 10000 picture elements (pixels) in its visual field (a pixel is equivalent to one of the dots in your TV screen). Experiments utilizing pixelized goggles suggest that as few as 625 pixels is sufficient for trained subjects to read text or navigate mazes.

Joseph Rizzo, of Harvard Medical School, was one of the first to believe that a visual prosthesis was possible. Trained both in neurology and ophthalmology, Rizzo says that he was in the perfect position to recognize the importance of a singular insight – in degenerative diseases of the eye that lead to blindness it is only the photoreceptor cells, the rods and cones, that are damaged. The rest of the electrical circuitry leading from the eye to the brain still works. It just lacks stimulation. Therefore, Rizzo reasoned that the rods and cones could be replaced with electronic versions thereof. In this way, most of the visual processing done by the brain could be retained.

Rizzo, who is from Harvard Medical School, and John Wyatt, an electrical engineer from the Massachusetts Institute of Technology, cofounded the Retinal Implant Project to develop a practical prototype from Rizzo's idea. In its current implementation, the retinal implant contains a photodiode array mounted inside of the eye on the front of the retina, opposite from where the normal visual image is focused. The photodiodes are stimulated not from the outside world directly but by an amplitude-modulated laser beam coming from a miniature camera mounted on a pair of spectacles. The laser beam also supplies the power to the electrical components in the eye (less than a quarter of a milliwatt is required for the current model). Output from the photodiodes would be circuited through a "stimulator chip" which would direct current into electrodes connecting directly to the ganglion cells, a million of which are the source of all visual input to the brain.

"It would do incalculable damage to the project," Rizzo warns, "if your story sensationalized or misrepresented the work done here. In particular, I would not want to falsely raise the hopes of the blind that a solution was near at hand."

The quality of vision supplied by the retinal implant is hard to gauge prior to human experimentation. Initially, it might be little greater than simple pattern recognition. In time, however, the simple photodiode array in the retinal implant might be supplanted by a "neuromorphic silicon retina" being developed at the California Institute of Technology, among other places. Neuromorphic means simply a design that mimics nerve cells. The beneficiaries of this "vision chip", so far, have been machines.

Machine vision has always been an intractable problem – despite myriad advances in digital signal processing, it still is difficult to design a robot that "sees" well enough to navigate through a cluttered room. It will be a long time before a mechanical "orb collection device" tries out for the major leagues. Although electronic systems can be designed with virtually unlimited resolution, how does the robot's digital brain know how to connect the dots, to make meaningful shapes and contours out of the pixelated input?

Biological vision utilizes massively parallel analog processors, usually called nerve cells, which combine output to extract from the visual field such details as edges, local contrast, and movement. Odd as it may sound, we do not see the world as it really is. Using only a fragment of the total electromagnetic spectrum available, from reflected wavelengths we impute to objects completely subjective properties like color and texture. Unlike a digital camera, our pixels "talk to each other." Long before the visual input gets to our brain, the data have been processed to make edges edgier, and contrast more visible. Movement is given special attention, especially in our peripheral vision. Neuromorphic chips try to implement this biological signal processing in silicon.

The advantage of designing neuromorphic chips according to Tobi Delbruck, who worked with Koch at Caltech, "... is that it forces you to adopt an efficient solution, in terms of processing power and silicon. The brain doesn't want to use a lot of wire, either." In addition to his exceptional repertoire of card-tricks, Delbruck, the son of Nobel prize winner Max Delbruck, is known for designing state-of-the-art "adaptive photoreceptors." Unlike a camera that whites out the image in the presence of too much light, the eye is capable of adjusting to logarithmic changes in light intensity. So too with Delbruck's neuromorphic photoreceptors, which use silicon transistors to accomplish the same thing.

In the silicon retina, logarithmic photoadaptors are connected to a two-dimensional analog grid, using design principles of the natural retina. Like biological systems, the silicon retina can sharpen edges and enhance features that are in shadow. Though the resolution of silicon retinas is still relatively low, it is perhaps sufficient for security applications such as recognizing faces or fingerprints, according to Christof Koch. Ultimately, he believes, "neuromorphic systems can provide a 'natural' substitute for damaged parts of the human nervous system, such as the retina ..."

Another sort of artificial retina, the ASR microchip developed by Alan and Vincent Chow, founders of Optobionics, is a silicon chip 2 mm in diameter and 25 μm thick that contains approximately 5000 microscopic solar cells called "microphotodiodes," each with its own stimulating electrode. These microphotodiodes are designed to convert the light energy from images into electrical/chemical impulses that stimulate the remaining functional cells of the retina in patients with age-related macular degeneration, retinitis pigmentosum, and similar conditions.

The ASR microchip is powered solely by incident light and does not require the use of external wires or batteries. When surgically implanted in the subretinal space, the ASR chip is designed to produce visual signals similar to those produced by the photoreceptor layer. Artificial "photoelectric" signals from the ASR microchip should induce biological visual signals in the remaining functional retinal cells, which may be processed and sent via the optic nerve to the brain.

In preclinical laboratory testing, animal models implanted with the ASRs responded to light stimuli with retinal electrical signals (ERGs) and sometimes brain-wave signals. The induction of these biological signals by the ASR chip indicated that visual responses had occurred.

The ASR microchip has had limited testing in humans since the year 2000. Twenty patients who suffer from retinitis pigmentosum, a degenerative disease of the eye, have been implanted. In one study, patients subjectively reported "improved perception of brightness, contrast, color, movement, shape, resolution, and visual field size [4]." Some patients were able to identify letters on eye charts, whereas they could not prior to the implantation. There were no safety issues identified.

Five-thousand pixels, whilst enough to allow a visual sense, is far from the type of vision the rest of us enjoy. The eventual commercial version of the artificial retina will probably involve a much larger array of photodiodes, achievable only through nanofabrication.

Optobionics is collaborating with the Hines Veterans Administration Medical Center, the Louisiana State University Eye Center, the University of Illinois Eye

Center in Chicago, Stanford University's Nano Fabrication Facility, and Tulane University Medical Center, to further improve the biocompatibility and function of the artificial retina microchip.

The Department of Energy (DOE) has committed over $9 million for artificial retina research. The money is being used, in part, to support research at the Doheney Eye Institute on the USC campus and at North Carolina State University as well as national laboratories.

The device favored by the Doheney Eye Institute is a miniature MEMS (microelectromechanical system) -type disc containing an RF receiver and an electrode array that can be implanted in the back of the eye to replace a damaged retina (Fig. 43). Visual signals would be captured by a small video camera in the eyeglasses of the blind person and processed through a microcomputer worn on a belt. The signals are transmitted to the electrode array in the eye. The array stimulates optical nerves, which then carry a signal to the brain.

The prototype implants contain 16 electrodes, allowing patients to detect the presence or absence of light. Six patients have been implanted. The patients are able to detect motion and in some cases, to identify objects. One patient who had been blind for fifty years was able to distinguish between a cup, a plate and a knife. The artificial retina project's "next generation" device would have 1000 electrodes which, hopefully, would allow the user to see images.

The project started with work at Johns Hopkins University under medical doctor and researcher Mark Humayun. When Humayun began the Intraocular Retinal Prosthesis Group at Doheney Retina Institute at the University of Southern California, the project moved with him. Teaming with Eli Greenbaum at Oak

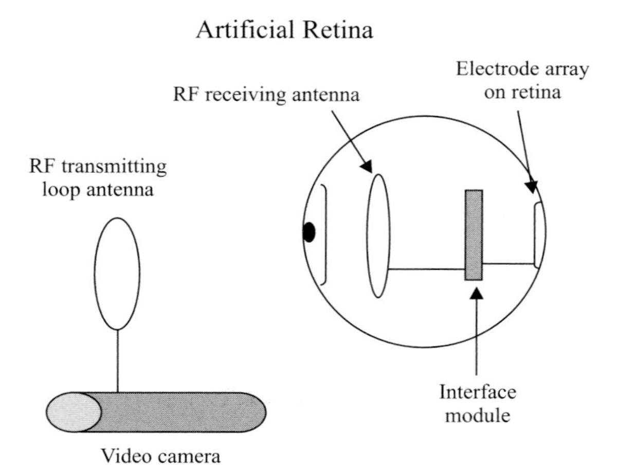

Figure 43 Diagram of the Doheney Eye Institute artificial retina. A video camera is mounted upon a pair of eyeglasses. Information from the video camera is transmitted via radio-frequencies (RF) to a an intra-ocular receiving antenna. The information is decoded by the interface module and transmitted via a direct connection to an array of electrodes contacting optic nerve endings in the retina.

Ridge, the pair is coordinating the project through a number of national laboratories.

The lead laboratory, Oak Ridge National Laboratory, manages the multi-laboratory effort as well as test the various components developed by the other labs. Argonne National Laboratory will investigate the viability of diamond-based electrode arrays and biocompatible coatings. Lawrence Livermore National Laboratory is experimenting with rubberized electrode arrays, while Los Alamos National Laboratory will model and simulate neural paths of and from the retina to the brain.

The retina cannot handle much pressure. Thus, the Sandia Labs approach favors spring-loaded electrodes that insure good electrode contact with minimal force. Also, protein fouling can mess up delicate interfaces intended to transmit electrical impulses. Other problems include biocompatibility and long-term reliability. The project, which has been under way since October 2001, is expected to identify the most promising implantation technologies.

USC personnel implant the devices and test their medical effectiveness. North Carolina State University in Raleigh leads the development of the *in-situ* medical electronics California-based Second Sight, is producing the implant, and hopes to commercialize the finished system. The Chairman of Second Sight is Alfred Mann, also the chairman of Advanced Bionics, which manufactures cochlear implants, as well as the chairman of the Alfred Mann Foundation, and a major benefactor of the Alfred Mann Institute for Biomedical Engineering at the University of Southern California. Mann is THE MAN when it comes to artificial organ systems.

Brain–Machine Interfaces

Recent advances in nanotechnology could allow the development of neuroprosthetic devices that link between neuronal tissue and mechanical devices, as for instance, artificial legs for amputees.

Microdevices for cell-electrode interfacing for both cardiac and neural cells have been available for *in-vitro* applications for many years. Some microarray-type devices have been implanted in rudimentary artificial vision systems, as discussed in the previous section. On a slightly larger scale, electrode systems, such as Medtronic's Activa, have been implanted in the brain to provide electrical signal stimulus that alleviate the symptoms of Parkinson's disease. These devices input information (or perhaps, disinformation) into the cells, but do not receive output. Another company, Cyberonics, has developed a similar device that impacts the vagus nerve. This was initially conceived as a therapy for victims of epilepsy, but clinical testing revealed a much wider market since stimulating the vagus nerves apparently relieves symptoms of extreme depression in some people, eliminating the need for antidepressant drugs.

Devices including "neuroprosthetic" limbs for paralyzed people and "neurorobots" controlled by brain signals from human operators would be the ultimate applications of brain–machine interface technologies developed under a $26 million contract to Duke University sponsored by DARPA. Such a device would have to translate brain activity into appropriate commands to move the prosthesis.

The contract is part of DARPA's Brain–Machine Interfaces Program (www.darpa.mil/dso/thrust/sp/bmi.htm), which seeks to develop new technologies for augmenting human performance by accessing the brain in real time and integrating the information into external devices. Besides the development of brain-controlled prosthetic limbs, neurosurgeons could apply brain-mapping enabled by the new technologies to aid surgeons in distinguishing healthy brain tissue from that which is part of a tumor or a focus for epileptic seizures.

Beyond medical uses, brain–machine interfaces also could be applied to enhance the abilities of normal humans, said the researchers. DARPA has initiated a program to build interfaces with the human brain that would a allow a person, using only his or her mind, to control peripheral devices and systems. Some of the aims of this protein involve: (1) scaling of position and motion such that a "slave actuator" could operate in workspaces either far larger (like a crane) or far smaller (like nanotech spaces) than our normal reach; (2) scaling of forces and power so that extremely delicate tasks (e.g., surgery) or high-force activities (lifting a truck) could be accomplished; and (3) scaling in time, so that tasks could be accomplished faster than normal human reaction time, for example, operating military aircraft. Neurally controlled robots could enable remote search-and-rescue operations or exploration of hazardous or inaccessible environments.

In 2000, Miguel Nicolelis and his colleagues at the Duke Center of Neuroengineering tested a neural system on monkeys that enabled the animals to use their brain signals, as detected by implanted electrodes, to control a robot arm to reach for a piece of food. The scientists even transmitted the brain signals over the Internet, remotely controlling a robot arm 600 miles away. The technique they used, called "multi-neuron population recordings" was originally developed by center collaborator John Chapin of the State University of New York in Brooklyn.

The scientists used arrays of up to 96 electrodes to sense signals from multiple areas of the brain, including the motor cortex from which movement is controlled. The researchers then recorded the output of these electrodes as the animals learned "reaching tasks," including reaching for small pieces of food.

The scientists fed the mass of neural signal data generated during many repetitions of these tasks into a computer, which analyzed the brain signals to detect tell-tale patterns that would enable researchers to predict the trajectory of the monkey's hand from the signals. Then, by programming the computer connected to the robotic arm to sense these signal patterns emanating from the monkey's brain, the scientists could enable the monkey to, in effect, control the arm only via neural signals.

This proof-of-concept experiment showed the effectiveness of recording from multiple areas of the brain and then allowing the computer to "learn" brain signal patterns that triggered certain movements.

Nicolelis, Craig, Henriquez and their colleagues aim to increase the number of recording electrodes to more than 1000 to enable control of more complex actions by robotic arms and other devices. The "neurochip" being developed by Patrick Wolf at Duke University will greatly reduce the size of the circuitry required for sampling and analysis of brain signals.

"Our dream is to develop a palmtop-like device that routes the signals either to robotic devices, computers, or even to the physician, to alert the physician to some problem," says Nicolelis.

Nicolelis, Henriquez and their colleagues are among researchers who believe that neurons are not hard-wired circuit elements permanently assigned to one computing task, like the microprocessor inside a computer. Rather, their theory holds that neurons are adaptable entities that can participate in many processing tasks at once. Moreover, these tasks may change from millisecond to millisecond. For example, Nicolelis' experiments have revealed that the brain signals producing a single event, such as a monkey reaching out, are mirrored in many places in the same brain region – as if the neurons "vote" on such actions.

In their current experiments, the center's scientists and engineers are developing "closed-loop" systems in which movement of the robot arm generates tactile feedback signals in the form of pressure on the animals' skin. Also, they are providing visual feedback by allowing the animal to watch the movement of the arm. Such feedback studies could also potentially improve the ability of paralyzed people to use such a brain–machine interface to control prosthetic appendages.

"One provocative, and controversial, question is whether the brain can actually incorporate a machine as part of the neural representation of the body," says Nicolelis. "I truly believe that it is possible. The brain is continuously learning and adapting, and previous studies have shown that the body representation in the brain is dynamic. So, if you created a closed feedback loop in which the brain controls a device and the device provides feedback to the brain, I would predict that as people or animals learn to use the device, their brains will basically dedicate neuronal space to represent that device." And so, we will be equipped with peripherals. Maybe we'll be able to print out our thoughts?

At least two private companies, Cyberkinetics Neurotechnology Systems, of Foxboro, MA and Neural Signals of Atlanta, GA, are working on brain interface technology. The immediate goal of these devices is to allow a victim of paralysis control a computer through which he or she could read, learn, and entertain him/herself, as well as communicate with the world via e-mail, etc. These devices are relatively simple and don't yet involve nanotechnology.

Cyberkinetics' BrainGate system is currently in clinical trials. This systems consists of two parts: an array of about 100 electrodes, each smaller than a human hair, that are implanted directly into the brain. These electrodes sense nerve impulses (action potentials) in individual neurons, and a neural signal processor that works to translate these nerve impulses into directed action. By a series of biofeedback training exercises, a person is able to control the cursor on a computer simply by using their own thoughts. The first BrainGate device was implanted into a male quadriplegic in June 2004. With it, he reportedly has learned to check his e-mail and play video games. Longer term, the company hopes that people will be able to replicate keystrokes, so that writing is possible without picking out letters one-by-one with the cursor. For all you-do-it yourself neuro-hackers out there, the company sells multielectrode arrays as a research tool.

Neural Signals offers two options; one is an implantable device similar to that of Cyberkinetics. Another is a “skull screw” that collects brain-wave data similar to an electroencephalogram.

The ultimate goal of both companies is to allow the paralyzed to communicate not just to a computer but once again with their dormant muscles, so that they may get out of their chairs and walk. Co-ordinated movement is apt to require much larger electrode arrays, and perhaps nanofabrication.

At the MIT Bioinstrumentation Lab, Sylvain Martel, who we met earlier for his work on nanorobots, is in collaboration with Nicholas Hatsopoulos and John Donoghue at Brown University on an implantable electrode array as computer–brain interface. In their view, the development of a chronically implantable array is an important step towards the long-term goal of creating a neural prosthetic to be used by humans. By implanting these arrays into the motor cortex of humans with spinal injuries, it may be possible for these patients to control devices such as a cursor on a computer monitor or even their own limbs by activating neurons in the motor cortex.

Towards this end, the team has already developed a statistical decoding algorithm that takes spike trains from multiple neurons that have been recorded in the past from a monkey's motor cortex and drives a simple robot arm in a direction consistent with that of the monkey's actual arm (see *Science* 286 (5441), **1999**). This team is now working on developing a real-time system that will take spike trains and replicate both the direction and the detailed trajectory of the monkey's hand as it is being performed.

James Baker, at the University of Michigan, is developing nanoscale biosensors and bioactuators for use in individual health and safety monitoring, in conjunction with external analytic bioNEMS (Nano-Electronic-Mechanical System) devices. This involves nanoscale polymer structures less than 20 nm in diameter as the basis of the sensor/actuators. The structures would be designed to target into specific cells of an individual and be able to monitor health issues such as the exposure to radiation or infectious agents.

“These bioNEMS,” says Baker, “would also be able to administer therapeutics in response to the needs of the individual, and act as actuators to remotely manipulate a person as necessary to ensure their safety.” One long-term possibility that he cites would be to activate muscle movement causing an unconscious person to walk out of harm's way, obviously a high level function.

A newly developed nanophase polymer surface could improve the interface between electronic implants and living tissue. David C. Martin, Director of the Macromolecular Science and Engineering Center at the University of Michigan, has presented research on polymers that can be processed into a “fuzzy” form to enhance the compatibility of electronic implants with brain tissue.

Electrodes implanted in the brain can pick up electrical signals sent back and forth by nerve cells. These devices are coated with growth factors that encourage brain tissue to grow into them. The intent is for each probe to make contact with a series of neurons, allowing it to receive signals it can interpret and use to activate an external device. One application for this technique is a spinal cord bypass in paralyzed patients. Patients with brain disorders and paralysis operate artificial limbs or control a computer mouse simply by thinking about the task.

Initial experiments in guinea pigs showed that these electrodes do not make efficient contact with the brain. "The implanted electrodes are solid, hard and smooth," Martin said, "whereas the brain is soft, wet and alive." The differences can cause the electrodes to lose contact with the brain, blocking the signal.

Martin and his team have designed rough-surfaced, fuzzy polymers with grooves and depressions designed to mesh better with neurons. "The scheme is to have these electrodes make a connection with the neurons quickly, before the other cells get in and wall them off," Martin says.

To further encourage connection, Martin and his team have incorporated biological molecules in the polymer coating to selectively attract target neurons. In guinea pigs, the researchers found that uncoated electrodes came out clean after remaining in the brain for a period of time, whereas coated electrodes were covered with neural tissue. This indicates that the neurons are hanging on to the biologically doped coating. The fuzzy surface of the polymer coating, in addition to improving contact with brain tissue, could be used to fine-tune its ability to conduct electrical signals.

From a medical standpoint, the most useful brain–machine interface would be one that allowed a paralyzed person to bypass his or her damaged spinal cord. Such a device would perhaps translate brain activity into RF signals that could be received by artificial motor neurons, that could activate the patient's muscles. Such artificial neurons, called Bions, are under development at USC (see next section).

Another device under development at USC is the artificial hippocampus, the part of the brain involved in memory. There are rats running around the Alfred Mann Institute as USC with such chips in their brain. Such a device might have a very large market based on aging adults who want to preserve their lifelong store of memories before inevitable neural degeneration erases them. It is difficult to relate rodent requirements for memory to that of a human being, however, and this device will require a long development course.

Artificial Cells

Gerald Loeb and his research group at USC are now working on bionic neurons (Bions). Bions are functional replacement for nerves that innervate muscles. They are RF-controlled devices small enough to be injected into paralyzed muscles, where they receive data and control sequences by radio links with an external controller. Computer models based on experimental data from muscles, motor neurons and proprioceptors are being developed to test new theories of control that may permit the reanimation of paralyzed limbs via functional electrical stimulation (FES). Loeb and his group have a collaborative relationship with Advanced Bionics, a company which makes cochlear implants. The Bion would take over the function of a motor neuron, without being in any way biological, however.

Robert Freitas, a research scientist associated with Zyvex and author of *Nanomedicine,* has put forth several imaginative designs for several artificial cells: respirocytes (an artificial red blood cell); clottocytes (an artificial platelet); and microbivores (artificial macrophages). Freitas' designs are of the "Fantastic Voyage" nanosubmarine variety, but lacking the miniature Raquel Welch to pilot them. These designs are not currently practical, due to the lack of a mode of manufacture. Presumably, a programmable assembler will be necessary before these designs can be realized.

The respirocyte would be micron-sized diamondoid storage tanks for transporting respiratory gases throughout the human body, which could be reversibly pressurized up to 1000 atm in direct response to changing tissue requirements. Molecular sorting devices powered by glucose oxidation would accomplish gas exchange.

The clottocyte is conceived as a serum oxyglucose-powered spherical nanorobot about 2 μm in diameter containing a fiber mesh that is compactly folded onboard. Upon command from its control computer, the device promptly unfurls its mesh packet in the immediate vicinity of an injured blood vessel – following, say, a cut through the skin. Soluble thin films coating certain parts of the mesh dissolve upon contact with plasma water, revealing sticky sections that adhere to blood group antigens. Blood cells would supposedly be trapped in the overlapping artificial nettings released by multiple neighboring activated clottocytes, and bleeding would halt at once.

The microbivore would be an oblate spheroidal nanomedical device consisting of 610 billion precisely arranged structural atoms plus another ~150 billion mostly gas or water molecules when fully loaded. This proposed nanorobot measures 3.4 μm in diameter along its major axis, and 2.0 μm in diameter along its minor axis, ensuring ready passage through even the narrowest of human capillaries (~4 μm in diameter). The microbivore has a dry mass of 12.2 picograms.

The microbivore would contain two internal materials-processing chambers totaling 4 μm^3 in displaced volume. The nanodevice would consume 100–200 pW of continuous power while in operation, and could completely digest trapped microbes at a maximum throughput of 2 μm^3 per 30-second cycle – large enough to internalize a single microbe from virtually any species in a single gulp, says Freitas.

During each cycle of operation, the target bacterium is bound to the surface of the microbivore like a fly on flypaper, via species-specific reversible binding sites similar to a cell-surface antibody. Telescoping robotic grapples emerge from silos in the device surface, establish secure anchorage to the microbe's plasma membrane, and then transport the pathogen to an "ingestion port" at the front of the device where the pathogen cell is internalized into a 2 μm^3 morcellation chamber.

After a thorough mechanical mincing, the remains of the bacterium are pistoned into a digestion chamber where a sequence of 40 enzymes would be successively injected and extracted six times, ultimately reducing the material to amino acids, mononucleotides, glycerol, free fatty acids, and sugars. These simple molecules would then be harmlessly discharged back into the bloodstream through an "exhaust port" at the rear of the device, completing a 30-second digestion cycle. Freitas claims that his microbivore might be superior to natural macrophages in that only inactive simple molecules would be released, whereas the real cell releases active fragments of proteins and carbohydrates that may cause hormonal or inflammatory responses.

Although Freitas' designs cannot be taken seriously, at present, he has done some careful thinking about how artificial cells might operate, and what sorts of constraints must be overcome to create synthetic replacements for cells. It is hard to imagine, however, that one could design a synthetic replacement that is better than the original cell. It seems more likely that cell therapies from transfusions or stem cell replacements will be more successful than artificially manufactured cells for most currently foreseeable purposes.

Re-Inventing Biology

Glen Evans, at founder of Egea Biosciences (now part of Johnson and Johnson), and others, such as Celera founder Craig Venter, have popularized the notion of "artificial organisms." While it is possible to imagine a man-made molecular assembler crossing the Darwinian Threshold, to become a synthetic life form, it is much easier and more practical to reverse engineer an already existing organism.

The idea is to "de-evolve" a bacterium or other single-celled organism to the minimal number of genes necessary to support life. Craig Venter, famous for leading Celera's successful Human Genome Project, has joined forces with Nobel prize winner Hamilton Smith, to create a synthetic genome, by starting with a simple mycoplasma genome, and eliminating genes not specifically required for replication and metabolism. Such a captive bug could then be reprogrammed to make bio-pharmaceuticals or other desirable products with high efficiency.

Robert Freitas, in his book *Nanomedicine*, Vol. I, points out that living amoebas have been reconstituted by the combination of five separate isolated fractions. Therefore, living organisms can be assembled from non-living components, verifying the unsettling idea that living creatures are just animated machines. Molecular biologists have, for years, reassembled viruses from separated components for various purposes.

Outputs from synthetic organisms could include biochemicals not found in nature. Proteins, for instance, are restricted to about 20 amino acids, although these are often altered after they become part of the protein chain. There are at least 22 amino acids that are used in natural systems; some organisms have an addition transfer RNA (tRNA) that interacts with one of the normal stop codons and transfers an amino acid to the polypeptide chain. Some researchers are trying to extend the repertoire further by creating artificial tRNA and amino acyl-tRNA synthetases charged with non-natural amino acids to create proteins with novel properties that could not exist in nature.

"Nanotechnological design and manufacturing may take advantage of the system of manufacture of proteins or it may use other approaches," notes Yaneer

Bar-Yam of the New England Complex Systems Institute. "Either way, the key insights of how proteins work shows the importance of understanding various forms of description (DNA), self-reproduction of the manufacturing equipment (DNA replication by polymerase chain reaction, or cell replication), rapid template-based manufacture (RNA transcription to an amino-acid chain), self-organization into functional form (protein folding) and evolutionary adaptation through replication (mutation of DNA and selection of protein function), and modular construction (protein complexes)."

Companies such as Applied Molecular Evolution and Genencor, in fact, are even now designing and optimizing enzymes and antibodies for uses that Nature presumably never intended.

Together, nanotechnology and biotechnology offer an impressive parts list from which to make useful devices. As Steven S. Smith and his colleagues have pointed out, these items can be made "addressable" by the incorporation of DNA or RNA nucleotides. "Molecular motors, DNA-based switches, DNA-based oscillators, enzymes, ribozymes, deoxyribozymes, gold particles, chromophores, fluorescence-quenching agents, antibodies, aptamers, and nucleic acid-binding proteins can all be ordered along nucleic acid scaffolds. The potential for construction of useful devices using the extraordinary wealth of functionality made possible by ordering these elements is quite broad," says Smith [3].

References

1 Gibson, W. *Neuromancer.* Ace Books, **1984**.

3 Chow, A.Y., Chow, V.Y., Packo, K.H., Pollack, J.S., Peyman, G.A., Schuchard, R. *Arch. Ophthalmol.* 122: 460–469 (**2004**).

3 Smith, S.. Addressable nucleoprotein assemblies. In: Nalwa, H. (ed.) *The Encyclopedia of Nanoscience and Nanotechnology.* American Scientific Publishers, Stevenson Randy CA (**2002**).

Chapter 9
Financing Nanotech Dreams

"Nanotechnology is the design of very tiny platforms upon which to raise enormous amounts of money," according to a definition favored by Lita Nelson, head of MIT's Technology Licensing Office. If you have devised a nanoscale platform, there are two basic sources of money for your venture: private and public. Generally speaking, the federal government of the U.S. does not give money directly to corporations but there are exceptions to this rule: DARPA, the research arm of the defense department, has been particularly active in financing some of the more speculative uses of nanotechnology. But, in general, DARPA uses fixed-term grants that cannot be renewed. Most of the billions coming out of the National Nanotech Initiative and the subsequent $3.7 billion bill passed in 2004 will go to the construction of infrastructure or the support of academic research. Only about 5 % will find its way into the corporate sector. In this chapter, we will first look at how entrepreneurial companies get money to finance their nanotech dreams, and then examine how the largesse liberated through the National Nanotech Initiative is being distributed.

Lita Nelson's office is in charge of translating MIT's research findings into commercial opportunities. She either licenses patents that the university holds to technology companies, or in some cases, she works with venture capitalists or "angel investors" to establish new companies, often with an MIT professor as the CEO or Chief Technology Officer. "Angel investors" are people, often retired executives, fortunate enough to have more money than they need who are willing to invest it in early-stage companies. Though these angels sometimes earn enormous returns on their investments, they are frequently motivated as much by an interest in technology and the process of venture formation.

Charlie Harris, Venture Capitalist

One of the venture capitalists that Ms. Nelson deals with on a regular basis is Charlie Harris, who keeps his connections to MIT solid by being a "life-sustaining fellow" of the university and a shareholder in its Entrepreneurship Center.

The Nanotech Pioneers. Steven A. Edwards

ISBN: 3-527-31290-0

Charlie Harris is a savvy technology investor and as it happens, the head of Harris & Harris, a publicly traded venture capital firm. After years of searching, placing bets on everything from software to decorative tiles, he has found finally happiness in one very high-tech area. Charlie had discerned signs and portents, heard rumblings and grumblings of this vast new field, nanotechnology, that would change the way we do – well, everything. This new technology would allow us to manufacture things from the atomic level up, with an elegance, grace and functionality that would make all of our current stuff look like relics of the Iron Age.

Figure 44 Charlie Harris at BCC's NanoBio Convergence 2004 Conference. Photo by Sally Edwards.

Despite being tiny, nanotechnology does not come cheap. "Even when you're dealing with a very early stage company in nanotechnology, we find that the budget for intellectual property work, the annual budget, may range between $250 000 to $2 million a year, which is a lot for a fledgling company," points out Harris. A small number of far-seeing firms concentrate on almost nothing but nanotech, like Harris & Harris, Lux Capital, and Millennium Materials Technologies Fund; a few others like Ardesta, Polaris, and Draper, Fisher and Jurvetson have a strong concentration on nanotech within their portfolios. According to

Business Week, more than $1 billion has now been invested in nanotech start-ups, most of it in the last year or two.

Many venture capital firms still eschew investments in nanotechnology. According to Lux Research, despite all the interest surrounding nanotech, venture capital investments in this field have actually declined, from $385 million in 2003 to about $200 million in 2004, accounting for only about 2 % of total funding.

One reason why the venture capitalists are nervous is the excessive hype that attends the "nanotech revolution" – they still haven't recovered the dot-com implosion of the late 1990s, and don't want to get caught in another bubble. Another problem is that potential liability issues exist – toxicology data for most categories of nanoparticles is non-existent, so the risk really isn't quantifiable. One major industrial accident involving nanoparticles, it is feared, would set the whole field back for years.

The main reason, though, that venture capital firms are hesitant to invest in nanotechnology is the absence so far of a viable "exit strategy." Venture firms typically have about a five-year time horizon for their investments. They are able to exit their investment, hopefully with inflated returns, when the entrepreneurial firm has been sold to a larger one, or has come public through an initial public offering, giving the venture firms tradable shares. The Initial Public Offering (IPO) market, however, which is so necessary to maintain and enhance the pool of venture capital available for speculative enterprises, has not yet fully recovered from the dot.com debacle at the turn of the millennium and the subsequent corporate scandals. In particular, there has yet to be an initial public offering from a nanotechnology-focused firm since that time. Until the IPO window opens, venture capitalists will continue to be conservative in their financing of nanotech firms. This leaves the field wide open for the more adventurous, like Harris & Harris. They estimate that the number of "tiny technology" companies (which includes some microscale companies involved in MEMS) was about 650 worldwide by 2004 and growing rapidly.

Unfortunately for Charlie Harris, nanotechnology represents a confluence of physics, chemistry, biology, and engineering – all subjects which Charlie, a Princeton English Lit major, had studiously avoided in college. While others were computing differential equations, he was deconstructing Moby Dick. While others were interpreting the sequences of DNA, he was reinterpreting Chaucerian syntax. While others carefully plotted the dance of electrons through pi orbitals, he was reveling in the lyricism of the Immortal Bard. So what to do?

Harris did what any great investor, from Benjamin Graham to Warren Buffet, would have done. He hired experts. Charlie's latest hire is Daniel Wolfe, out of the Harvard lab of George Whitesides, a major nanotech star. Other members of the team include Daniel Leff, who received his Ph.D. in Physical Chemistry in the CalTech laboratory of nanotech pioneer James Heath; Kelley Kirkpatrick, who co-authored the National Nanotech Initiative; Doug Jamison, who worked in the Technology Transfer office at the University of Utah; and Lori Pressman, former Director of the Technology Transfer Office at MIT.

Fortunately for Harris, in addition to his English Lit degree, he also sports an MBA from Columbia, so he already knew a little bit about finance.

Technology, it is said, transfers best in those objects that wear shoes. Charlie invests heavily in those sorts of objects; some of his investments have been in companies with one employee. For instance, when Charlie first discovered it, Neurometrix, consisted only of Shai Gozani, a Harvard M.D. Now, Gozani is President and CEO of a company that just had its initial public offering, yielding about $24 million.

Venture capital is a risky business. According to Harris, of 38 investments that his company has so far closed out, 21 lost money. But he made a lot more money on the winners than he lost on the losers. He and his company have grown $40 million in investments into a $108 million return with an average holding period of 2.7 years, for better than a 40% average annual return on equity. And that doesn't include the likely profits from the Neurometrix IPO, or the possible initial public offering of one of his premier nanotech portfolio companies, Nanosys.

The exclusive focus on tiny technology, which Harris defines as "microsystems, MEMS (microelectronic mechanical systems) and nanotechnology" is relatively recent for Harris & Harris, dating back only to 2002. "There are three things you need for a successful venture capital company," says Harris, "Deal flow, talent to assess opportunities, and money to invest." By 2002, Harris had determined that there were a sufficient number of nanotech and MEMS companies to allow for a reasonable deal flow; of 700 such companies in his database, Harris has held discussions with about 160. And by 2002, the buzz around nanotech had grown to the extent that he could be assured that investors would be interested.

Nanotech companies already predominate in the H & H portfolio, including Nanosys, Nanogram, Nanotechnologies, Inc., Nanopharma, NanoOpto, Nantero, Molecular Imprints, Agile Materials & Technologies, Optiva and Neophotonics Corp. (Table 15). Other companies, like Chlorogen, Nanopharma, and Solazyme might be considered biotechnology companies, but their technology is also focused on the nanoscale.

Table 15 illustrates a cardinal point that Harris likes to make: an expertise in nanotechnology lets the venture capitalist diversify into a number of different industries. Harris' nanotech portfolio includes companies that operate in the semiconductor, communications, electronics, pharmaceutical, materials and energy fields. Even with a tiny company, nanotech can be employed in various ways. Cambrios, for instance, was established with intellectual property from the Angela Belcher lab, in which the self-assembly techniques of the natural world are employed to make industrial objects. Cambrios is most interested in the semiconductor field, but the same technology can be employed wherever a cheaply assembled material with nanoscale features would be desirable.

Venture capital firms need to have an exit strategy; most hope to achieve a return of capital within five years, hopefully with a return on capital as well. The two most common roads to a profitable investment are: (1) that the portfolio company is bought out by a larger company; or (2) that the portfolio company has an initial public offering; the venture company can either sell its holding in the offering, or take profits on the public market.

Table 15 Harris & Harris portfolio companies that work at the nanoscale.

Company	Industry	Focus
Cambrios	Semiconductor	Biomolecule-directed nanoscale material assembly
Chloragen	Pharmaceuticals	Protein production in plant chloroplasts
Molecular Imprints	Semiconductor	Nanolithography
NanoGram Corp.	Materials	Nanopowders and thin films
Nanomix	Semiconductor	Carbon nanotube and silicon devices
NanoOpto Corp.	Communications	Nanoscale optical components
Nanopharma	Pharmaceuticals	Nanoparticle drug delivery
Nanosys	Electronics	Nanowires, nanotubes, quantum dots
Nanotechnologies, Inc.	Materials	Nanopowders and coatings
Nantero	Semiconductor	Nanotube-based memory chips
NeoPhotonics	Communications	Planar optical devices using nanomaterials
Optiva	Electronic display	Nanomaterial-based flat panel displays
Solazyme	Energy	Photosynthetic microbes

Harris can afford to have a more long-term strategy than most. Since it is publicly traded, H & H has "permanent capital." Any investor who wants to can cash out his or her holdings simply by selling their stock. Nevertheless, Harris has already cashed out one of its nano investments. Nanogram Devices employs film technology to design a more efficient, longer-lasting battery for medical implants such as pacemakers. It was a spin-out from an H & H company, Neophotonics, which was itself spun out of Nanogram Corp. Nanogram Devices was swallowed up by Wilson Greatbatch, which has a virtual monopoly on the market for medical implant batteries.

Led by serial entrepreneur Larry Bock, one H & H portfolio company called Nanosys (see Chapter 7) raised some eyebrows with its own appetite for intellectual property. Nanosys has spent heavily on intellectual property related to self-assembling inorganic "nanowires" and "quantum dots" – exotic components for future nanochips. The company hopes to assemble these into "exquisitely sensitive chemical and biological sensors" and really, really small electronic devices.

In September 2004, Nanosys had planned to test the public's appetite for nanotech shares with a widely watched public offering. However, the IPO was withdrawn at the last minute with the company citing "market conditions."

Nantero (which is also discussed in Chapter 7) is building memory chips out of nanotubes (NRAMS), with the aim of eventually of replacing all extant types of computer memory, a $100 billion dollar market.

H & H has other portfolio companies aiming at the chip industry. NanoOpto is developing and manufacturing optical communications and optical drive subcomponents on a chip. Neophotonics is working on planar optical devices using thin-film technology. Optiva is commercializing nanomaterials for optical applications, aimed initially at the flat-panel display industry. Molecular Imprints has invented a "Step and Flash" lithography system that can replicate features as small as 20 nm; hopefully, this system will facilitate the next step mandated by Moore's famous law, whereby semiconductor chips get faster and smaller with each new generation.

Nanotech and biotech have been on a collision course for the past few years, and one of the results has been the H & H portfolio company Nanopharma, which finds medical uses for its proprietary polymer nanoparticles. The company's scientific founder, Russian scientist Mikhail Papasov, had the good fortune to work at Massachusetts General Hospital alongside David Elmaleh, described as the "spiritual founder" of Puretech Ventures. Puretech helps puts together life science companies to commercialize technology that is too early-stage to attract most venture capital. And when they do need capital, one of the people they call is Charles Harris.

Papasov has invented a polymer called Fleximer, a biological "stealth technology" that mimics some aspects of carbohydrate chemistry but does not attract the attention of the body's immune system. This polymer can be used as a drug delivery system to introduce pharmaceuticals into the body that might be insoluble or that might be rapidly cleared from the system on their own.

Oncology is Nanopharma's first focus, according to CEO Peter Leone. He says that the company's lead drug is a generic chemotherapy agent that has favorable pharmacokinetics when combined with Fleximer. He describes the drug formulation as looking like a catamaran, with the generic drug forming the hull, and the outriggers composed of Fleximer. These particles are just the right size, "big enough so that they are not eliminated by the kidney (the body's dialysis system), but small enough that they can pass through the blood vessels and into the lymph system and intercellular spaces." The chemical links between Fleximer and the drug provides for controlled release. Nanopharma has a total of five proprietary small-molecule compounds that it says have clinically proven efficacy against human cancer.

Fleximer can also be used to extend the half-life of protein therapeutics, of which there are about $23 billion sold annually, a market that is growing rapidly. Current technology for extending protein half-life involves the chemical addition of polyethylene glycol (aka antifreeze) to the protein chain. There is evidence that polyethylene glycol can accumulate in the body with repeated injections. Thus, the company believes that Fleximer might be a safer alternative.

Another technology that Nanopharma is working on is system for targeting drugs to lymph nodes, which the company says will have application for cancer, AIDS, and biodefense. Also, they have developed a "Nanocassette" that mimics natural biocarrier molecules for targeting drugs to the nucleus of the cell. This would be particularly valuable for chemotherapy drugs that are generally very insoluble.

Nanopharma has entered into collaborations with two major pharmaceutical companies and a smaller company that is focused on eye care.

H & H has invested in another biotech Chloragen, whose trick is to produce proteins inside the chloroplasts (the organelle responsible for photosynthesis) of plants. In this way, protein drugs can be harvested from tobacco fields. The advantage of Chloragen's technology is that by sequestering the protein-encoding genes inside the chloroplast, they are prevented from leaking into the environment through pollen. In theory, such transgenetic pollen could fertilize other plants, allowing foreign proteins to be produced where they are not wanted. With Chloragen's technology, you don't have to worry about smoking protein hormones along with your cigarettes.

So what is next for H & H? Recently, the company had a secondary offering of its stock that raised $36 million and will be to fund its continuing quest for "world domination," in the words of one nanotech blogger, Cientifica's Tim Harper. Mr. Harris modestly denied any such ambition. However, he has set his sights on the next wave of nanotech innovation.

The first generation of nanotech products has already hit the shelves: stain-resistant fabrics, invisible sun-blocking particles for cosmetics, nanoparticles for polishing semiconductors and fluorescent quantum dots for bioassays. Harris expects the next generation to include "active nanostructures" such as transistors, actuators, and drug-targeting devices. Portfolio companies at H & H are already working to make these a reality. In the longer term, the company expects that integrated nanosystems will be developed that allow high-density electronic devices with transistors composed of individual molecules, polymer superstructures that aid in the engineering of regenerated tissues and organs, and possibly even "quantum computing."

Implementing the National Nanotech Initiative

While industry is the primary financier of developing technology, funding basic research is the province of the government. The days of the dedicated scientific amateur, like Erasmus Darwin or Robert Boyle, are long gone. Nevertheless, we owe much to these periwigged men of the British Royal Society and similar organizations around Europe. They brought us, kicking and screaming, into the modern age. Whatever the delights of metaphysics and alchemy, adherence to a belief in the supernatural got us nothing but the Dark Ages. Simple, repetitive, organized observation using the five natural senses and handmade instruments did much to banish the darkness of superstition. By the twentieth century, though, the five senses were not enough. Scientists needed tools to work with (toys to play with) and they cost money. A good electron microscope, for instance, today might easily cost $100,000.

In America, science was really the forgotten stepchild of industry until the twentieth century. Thomas Edison, for instance, really knew precious little about how electricity actually works – the electric light was actually invented before the electron. The American chemical industry got its start in granny's lye soap. Petrochemicals came later. In the late nineteenth century, my great-great grandfather, Hiram Everest started the Vacuum Oil Co. in collaboration with an inventor friend of his who had invented a vacuum distillation process by which petroleum – that nasty, smelly stuff that was oozing out of the ground in Pennsylvania – could be converted into a lubricant, replacing whale oil. Hiram was not a chemist, but he knew a little about buying and selling. His Vacuum Oil Co. was the start of the petrochemical industry, which eventually made the internal combustion engine possible. In time, Hiram sold out his interest to John D. Rockefeller, who merged his companies to create Socony Vacuum Oil, which later became Mobil Oil.

American science got a bonanza in the years before and after World War II, when a flood of scientists emigrated from Europe. Some, like Albert Einstein were Jews, escaping Nazism. It was Einstein's famous letter to Franklin D. Roosevelt that was the impetus for the Manhattan Project, which resulted in the atomic bomb. If there was ever any doubt, the bomb convinced the U.S. government of the value of the scientific enterprise. Large-scale federal funding has been with us ever since.

Funding of nanoscience by the government took place on a more or less *ad hoc* basis, until Mike Roco got into the act (as recounted in Chapter 2). Before Roco, nanotechnology *per se* wasn't even on the radar screen. In fact, because of its association with Drexler and his followers, nanotechnology had a kind of hard-to-fund science-fiction aura. But give Drexler credit: he was largely responsible for giving nanotech a separate identity in the public consciousness. Before Drexler, nanotech was just a disparate group of disciplines that were interrelated on the basis of scale.

Table 16 lists some of the critical events in the establishment of the National Nanotech Initiative (NNI) and the eventual passage of 21st Century Nanotechnology Research and Development Act. Roco's original informal group was not started before 1996, after carbon nanotubes, buckminsterfullerene, quantum dots, and the atomic force microscopy had all been invented. The Interagency Working Group for Nanotechnology (IWGN) was formally established in 1998. Roco gave his fateful presentation to Clinton's advisors in the White House during March, 1999. That was the first real indication that large-scale government support for nanotechnology might be in the works. In 2001, the IWGN was formally replaced by a subcommittee of Nanoscale Science and Technology in the National Science and Technology Council. Roco was made chair of the subcommittee. The first separately budgeted support for nanotech began in 2001, under the newly elected administration of George W. Bush.

The NNI is controversial in some quarters: "The much-heralded US National Nanotechnology Initiative (NNI) has been criticized for using 'nano' as a convenient tag to attract funding for a whole range of new science and technologies." says Alexander Arnall [1]. "This reinvention is one way of attracting more money because politicians like to feel they are putting money into something new and exciting."

Table 16 The timeline of the National Nanotech Initiative.

Date	Event
November, 1996	Informal working group established by Mike Roco, which included Stan Williams, Paul Alisatvos and James Murray, among others.
September, 1998	Interagency Working Group on Nanotechnology (IWGN) established under the National Science and Technology Council
March, 1999	Presentation by Mike Roco at the White House to presidential economic and science advisors.
September, 1999	Publication of *Nanotech Science and Technology: A Worldwide Study* by IWGN
January, 2001	Bill Clinton formally announces the National Nanotech Initiative and included it as a federal initiative in the 2001 budget proposal
2001	IWGN replaced by Subcommittee on Nanoscale Science and Technology with Mike Roco as Chair
December 2003	21st Century Nanotechnology Research and Development Act passed by Congress and signed into law by George W. Bush

Although it is true that some of the projects being funded under the NNI could have been funded separately under initiatives that did not bear the identifying term "nanotechnology," this criticism seems to miss the whole point. Nanotechnology is exciting precisely because it unites a diverse group of technologies into a single discipline.

With apologies, I will here interpose a long analogy that would perhaps only occur to a biologist. The Swedish scientist Carl Linnaeus was one of the first to make a concerted attempt to describe and group the living things of this world. "The study of nature," said Linnaeus, "would reveal the Divine Order of God's creation." Although many of his classifications were, in retrospect, clearly wrong, it is from his work in the early eighteenth century that we get the species and genera of classical taxonomy. The living world as seen by Linnaeus was a complicated place. Tens of thousands of species have been described, with many times that many waiting to be discovered, and some that have not yet come into being.

It was not until the late twentieth century, after the molecular basis of genetics was unraveled, that it became obvious that the biochemical basis of all life on this planet was essentially identical. Darwin's "Theory" stopped being a theory and became an obvious, undeniable fact. At the level of the individual cell, bacteria, plants, fish, amphibians, mammals, etc., all organisms work pretty much the same (with the potential exception of nanobacteria). One could even say that there really is only one form of life, albeit with widely divergent manifestations.

In Linnaeus' time, all scientists were called Natural Philosophers, and there was little division as regards to disciplines. Though Isaac Newton, for instance, is known best to us as a mathematician, he had a serious interest in optics, physics, celestial astronomy, chemistry and metallurgy. No one suggested that he wasn't qualified for any of these pursuits; in fact they were regarded as part of the same pursuit – the identification of natural laws.

By the end of the twentieth century, science had ramified into thousands of branches. My Ph.D. is in the rather prosaic field called biology, but I have been variously called a virologist, a molecular biologist, a geneticist, developmental biologist, a protein biochemist and I held an appointment as an assistant professor of biochemistry.

At the level of the atom, all of these various disciplines become meaningless to a degree. Atoms will not combine into molecular clusters or crystals differently because they are coaxed to do so by a biologist, a chemist or a physicist. The whole point of nanotechnology is that one discipline will suffice. In essence, nanoscale completes the circle. All scientists can become natural philosophers once again.

The United States federal government, of course, is less concerned with the development of pure science than with basic research that can support the technology on which industry is built. Realizing this, Roco and the National Science Foundation put together a now famous estimate that nanotech could contribute $ 1 trillion to the world economy by 2015. A breakdown of that $ 1 trillion is shown in Table 17. About one-third comes through the development of new "nanomaterials." About 30 % is contributed by nanoscale computer chips and integrated circuits. It was estimated that nanotech-based pharmaceuticals would be amount to about half of the total, or about $180 billion. Chemical catalysts, aerospace, and nanotools make up the remainder.

Table 17 The one trillion dollar industry by 2015?

Industry	Annual revenue	Description
Nanomaterials	$340 billion	New materials and processing
Electronics	$300 billion	Semiconductors, integrated circuits
Pharmaceuticals	$180 billion	Nanotech will be used in the production of 50 % of pharmaceuticals by 2015
Chemicals	$100 billion	Nanostructured catalysts in chemical and petroleum processing
Aerospace	$70 billion	Aircraft skin, frames, propulsion
Tools	$22 billion	Measurement, imaging, simulation (scanning probe and electron microscopes, nanomanipulators molecular modeling)

The $ 1 trillion figure is certainly eye-catching, but one should not get the impression that nanotech is leading us, therefore, into an era of untold wealth and prosperity. Roco carefully avoids saying that nanotech would give us NEW industries worth a trillion dollars. Rather, old industries will adopt nanoscale technologies. Does it make sense to redefine the semiconductor industry as nanotechnology because the 100-nm barrier for chip architecture has been broken? How much of the $180 billion in pharmaceutical nanotech could just as easily be called biotech?

Roco and the NSF also say that about two million new workers will be needed to work in new nanotech areas by 2015. Arrayed against the one trillion figure, this is not such impressive a figure. If that one trillion were actually NEW production, it would mean that each new nanotech employee would be contributing $0.5 million to the economy, a pretty healthy level of productivity. But we are not really talking about new production; instead, it is the replacement of old technologies with new technologies – the usual creative destruction through which modern economies evolve. That does not mean that we shouldn't make the investment. The country or company that creates the technology first will likely reap the reward.

The President's Council of Advisors on Science and Technology (PCAST) recently completed a five-year assessment of the NNI. They concluded that it was money well spent and that, partly as a result, the United States is now the acknowledged world leader in nanotechnology when measured in a variety of ways: money spent, papers published, patents issued or applied for. This "leadership position, however, is under increasing competitive pressure from other nations as they ramp up their own programs," notes the Council.

The level of investment in the industrial sector is growing such that it will soon outweigh the government as a source of nanotech funding, if it hasn't already, according to the Council. One area where the U.S. has a real advantage over the rest of the world is in the creation of new technology companies. This country has a culture of entrepreneurship and a capital pool for new ventures that is really unmatched anywhere else.

Because nanotech can be applied to such a wide range of industries, a wide range of government agencies can help distribute the pie represented by nanotech funding. Part of the way that Roco and his associates were able to sell the NNI was to identify divisions of the federal government that might benefit. Table 18 illustrates the agencies involved, and it reads like an organizational chart of the executive branch of U.S. government. There is not much of importance left out except for the Department of Labor and the National Park Service. National security concerns are a big item, reflected in the inclusion of the Department of Defense, the Department of Homeland Security, and the intelligence community.

"Since its inception, the NNI has done a very good job of organizing the pertinent Federal Government agencies around the nanotechnology topic, establishing a robust national research infrastructure..." notes PCAST in its five-year review. "With twenty-two different participating agencies, each with its own distinct mission, these accomplishments deserve high praise."

Table 18 Federal Agencies that participate in the NNI.

Consumer Product Safety Commission
Department of Agriculture
Department of Commerce
Department of Defense
Department of Energy/Office of Basic Energy Sciences
Department of Health and Human Services (NIH, FDA, CDC, NIOSH)*
Department of Homeland Security
Department of Justice
Department of State
Department of the Treasury
Environmental Protection Agency
Intelligence Community (CIA, FBI, NSA)**
International Trade Commission
National Aeronautics and Space Administration
National Science Foundation
Nuclear Regulatory Commission
Patent and Trademark Office

* CDC: Centers for Disease Control; FDA: Food and Drug Administration; NIH: National Institutes of Health; NIOSH: National Institute of Occupational Safety and Health.
** CIA: Central Intelligence Agency; FBI: Federal Bureau of Investigation; NSA: National Security Agency.

The biggest recipients of funding under the 21st Century Nanotechnology Research and Development Act – the successor to the NNI – are the National Science Foundation, the Department of Energy (DoE), the National Institutes of Standards and Technology (part of the Department of Commerce) and the Environmental Protection Agency (see Table 19). It should be stated that a lot of the DoE funding has nothing to do with energy. A legacy of the Manhattan Project during World War II was the establishment of national laboratories in what were then out-of-the-way places, like Oak Ridge, Tennessee or Los Alamos, New Mexico. After the war, it was decided that a peaceful use for the splitting of the atom must be found; hence the national labs became the property of the DoE. Now, the DoE runs a chain of them, including not only Los Alamos and Oak Ridge, but also Brookhaven in New York, Argonne in Illinois, and Lawrence Livermore, in California. These labs still carry out energy research, but they also have active programs in biology, physics, materials research, and much more. The national laboratories have become one of the biggest beneficiaries of the drive toward nanotechnology.

Table 19 Budget for the 21st Century Nanotechnology Research and Development Act.

Agency*	FY 2005	FY 2006	FY 2007	FY 2008
NSF	385	424	449	476
DoE	317	347	380	415
NASA	34	37	40	42
NIST	68	75	80	84
EPA	5.5	6.05	6.41	6.8

* DoE: Department of Energy; EPA: Environmental Protection Agency; NASA: National Aeronautics and Space Administration; NIST: National Institute of Standards and Technology; NSF: National Science Foundation.

NASA is also a beneficiary of the nanotech dollar. The use of nanotech materials for airplanes and spacecraft is one large area of interest. Propulsion technology for spacecraft may turn out to be another. The development of a brain–computer interface is a Star Trek-type of project in which NASA purports to have a long-term interest. NASA has also become worried about nanobacteria because of the unexplained tendency for astronauts to develop kidney stones and other calcification problems after prolonged periods in space, which is possibly related to the growth of nanobacteria under zero-gravity conditions.

The National Institute of Standards and Technology (NIST), as the name would suggest, is charged with developing manufacturing standards, among other things. One of the major problems with nanotechnology is just coming up with accurate measurements. How do you come up with a usable nanometer-scale measuring device? How do you create a manufacturing technology where the tolerances are measured fractions of a nanometer? The NIST, in fact, has developed rulers that are precisely etched lines of crystalline silicon ranging in width from 40 nm to 275 nm. The spacing of atoms themselves within the silicon crystals is used in the way that hash marks on a school-boy's ruler to measure the dimensions of these test structures. Industry can use these rulers to calibrate their measurement tools. The semiconductor industry, especially, is one in which all the major players must be on the same page. They used to have five-year roadmaps; now they are starting to think 20 years ahead. The NIST helps to keep them together.

The Environmental Protection Agency gets a small piece of the nanotech pie, but it is an important piece. If there is anything that could derail the nanotech dream, it is the public perception that nanotech is dangerous. Improperly used, nanotech could almost certainly, in fact, be dangerous (this issue is discussed more thoroughly in Chapter 11, Fears of Nano). For this reason, Roco and his colleagues built discussion of societal and environmental impact into the NNI form the beginning. Part of the EPA money will be used to fund institutions, such as the Center for Biological & Environmental Nanotechnology at Rice University, where research is conducted on such problems as the toxicology of carbon nanotubes and other nanoparticles.

Mike Roco points out that one of the side benefits of the NNI has been a greater appreciation of the nanoparticles that are already in the environment. Asked to rate the hazards of nanoparticulates, Roco cites the products of imperfect combustion that we have always lived with without realizing it – carbon nanotubes and fullerenes among them.

The commercialization of nanotechnology actually goes back 1000 years or so if you include such nanomaterials as carbon black used in ink and paint, or the gold nanoparticles that give stained glass its color and lustrous appearance. Current nanomaterials, such as carbon nanotubes, dendrimers, quantum dots, or titanium dioxide nanopowders, have found their way into commercial products only in the past five years or so. The next generation of nanoproducts will be active structures, such as actuators, sensors, transistors, or memory devices. These have started to appear in prototype form in the year or so, and commercialization may start by the end of 2005.

The third generation is expected to consist of entire nanosystems, integrated on a chip, or encapsulated into a medical device. An example that Mike Roco described in one of his talks is the "nanoscale single-electron switching arrays for self-evolving neuromorphic networks." A single-electron switch is one that only needs to transfer a single electron to switch states – the ultimate in efficiency. A neuromorphic network is one that emulates neuronal systems, like our brains. A self-evolving neuromorphic network is a brain that can teach itself – like our own brains when we were still young. These are expected to start appearing around 2010.

Finally, molecular devices, including molecular transistors and other electronic components should begin to appear by the year 2020. One might expect commercially viable quantum computers shortly thereafter.

"Grand challenges" are usually specific goals, like the atom bomb, or the Apollo moon landing, that stretch the abilities of current technologies. The NNI has come up with some Grand Challenges for nanotechnology, as listed in Table 20.

These "Grand Challenges" are more like general programmatic areas of focus rather than specific goals. These are not Moon Landing sort of projects, and are therefore a little frustrating to the would-be chronicler of great technological events. For instance, Grand Challenge #1: Designed nanostructured materials. Mankind has, of course, always used nanostructured material, like wood, for instance (discussed in Chapter 1), or wool. The difference implied here is one of design. We didn't design wood or wool to be what it is; we just took advantage of it. The design of the next wool replacement, however important economically, is not the sort of inspiring project that is apt to launch a new generation of scientists.

Table 20 NNI "Grand Challenges."

1. Designed nanostructured materials
2. Nanoelectronics, optoelectronics, and magnetics
3. Nanoscale devices for therapeutics, and diagnostics
4. Environmental improvement
5. Efficient energy conversion and storage
6. Microcraft space exploration and industrialization
7. Chemical /Biological/Radiological/Explosive Threat (CBRE) detection and protection
8. Instrumentation and metrology
9. Improvement in manufacturing processes

Other challenges are equally nondescript. Environmental improvement. Great. We all want that, as long as it doesn't cost anything or prevent us from using our high-powered internal combustion engines But what does it mean exactly?

The National Nanotech Advisory Panel, a unit of the President's Council on Science and Technology "believe that nanotechnology is at too early a stage and too diverse to be pigeonholed into a few grand challenges." I respectfully disagree.

So in order to dramatize what can actually be done with nanotechnology, *Nanotech Pioneers* has made up its own (smaller) set of Grand Challenges, which we will address in Chapter 10.

Reference

1 Arnall, A.H., *Future Technologies, Today's Choices.* A report for the Greenpeace Environmental Trust, July 2003. www.greenpeace.org.uk.

Chapter 10
Mega-Sized Projects that Could Use Tiny Technology: Three Somewhat Grandiose Challenges

Massive, imaginative projects have the potential to inspire rapid advances in technology. Classic examples are the Manhattan project, which led to the atomic bomb, the Apollo project, which led to the first men on the moon and, more recently, the Human Genome Project, which led to the complete sequencing of the human genetic complement. Megaprojects such as these focus finance, project management, and scientific talent toward the realization of specific goals. Competition between research teams is every bit as fevered as that between football teams, and the result can be progress that is much more rapid than expected. In this chapter we will examine three "Grand Challenges" that involve nanotechnology: (1) Energy independence; (2) The Space Elevator: and (3) Quantum Computing. The first challenge admittedly owes a lot to one of the NNI Grand Challenges – Nanoscale Research for Energy Needs; the other two are brought to you solely courtesy of *Nanotech Pioneers.*

In announcing the Apollo project, President John F. Kennedy said that America would put a man on the moon within a decade. At *Nanotech Pioneers,* we are willing to be generous with the nascent nanotechnology industry, and double the time frame. If all three grand challenges are answered by the year 2025, we will be more than satisfied. All participating researchers will be thoroughly congratulated in the 20th anniversary revision of this manuscript (should Wiley VCH still be interested), and I will personally inscribe all copies purchased by said researchers. What a deal (my unsubstantiated research suggests that author-signed copies yield a 20 % greater return on E-bay's used-book sales)!

We need to give our challenges more concrete and understandable terms, to give participants firm goals to work for. So, by 2025, we expect that, in collaboration with other scientists and engineers, nanotechnologists will have:

- reduced world consumption of non-renewable carbon-based energy sources (petroleum, natural gas, and coal) by 50 %;
- constructed an elevator into outer-space, 62 000 miles above the Earth's surface; and
- manufactured an affordable and commercially available quantum computer.

The Nanotech Pioneers. Steven A. Edwards

ISBN: 3-527-31290-0

If technology advance is logarithmic, as Ray Kurzweil suggests, then the technology to achieve these goals should be well-in-hand by 2025. I would warn, however, that powerful economic interests protect the *status quo*, particularly in the area of energy. Likewise, the semiconductor industry may see the quantum computer as a destructive rather than disruptive technology. Technology, *per se*, may not be enough. One has to convince the world to use it.

Energy: Independence from Fossil Fuels

The United States of America obtains about half of its energy needs from overseas in the form of petroleum products. This has warped the nation's foreign policy, seriously damaged its balance of payments, and arguably cost the lives of more than 1800 American soldiers (and counting) and many times that number of Iraqis. Other industrialized nations, though less warlike than the U.S., are more or less in the same boat. Japan, for instance, has almost no energy resources. Rapidly developing nations such as India and China have aggravated the competition for energy. Even so, only a tiny proportion of the population of these two giants yet lives in a manner comparable to householders in Europe or the U.S. If economic growth continues in Asia the way that it has for another decade or more, it is scarcely possible to imagine where the energy will come from.

The over-reliance on carbon-based fuels is apparently responsible for the increase in carbon dioxide in the atmosphere and oceans, which is in turn causing a greenhouse effect increase in temperature – global warming. The increase in Asian consumption has made the modest and probably unattainable goals of the Kyoto Treaty in cutting the use of carbon fuel sources in the West completely inadequate to the task of forestalling global warming. The U.S. Department of Energy has determined that approximately 80 % of all human-caused carbon dioxide emissions currently come from fossil fuel combustion, and that world carbon dioxide emissions are projected to rise from 6.1 billion metric tons carbon equivalent in 1999 to 9.9 billion metric tons in 2020. The growth of human population carbon dioxide in the atmosphere and global warming seem to be inextricably linked.

One way to cut this linkage is to reduce the usage of carbon-based fuels, or possibly, to create the carbon fuels that we do use directly from the carbon dioxide in the atmosphere.

In the foreseeable future, nanotechnology is not likely to eliminate the use of petroleum-based energy source, but it is possible that we can seriously reduce consumption, in part by increasing the energy efficiency of the products we use, and in part by increasing the use of solar-based or possibly fusion-based energy. The goal of this energy challenge would be the reduction in the use of petroleum-based fuels by 50 % by 2025.

Energy is supplied to the inhabitants of Planet Earth in at least four different forms:

- Solar energy
- Gravitational energy
- Geothermal energy
- The atomic energy within matter.

Petroleum can be thought of as stored solar energy; oil is the organic residue of ancient photosynthetic plants. Some natural gas deposits may have other origins as methane was part of the atmosphere of the prebiotic Earth. Either way, these fuels are being used far more quickly than they can be replaced.

The world demand for carbon-based energy (oil, natural gas, coal, tar shale) is about 200 billion barrels oil equivalents. The demand is expected to rise to 300 billion barrels by 2100. Oil production will peak around 2020, and only by a heavy switch to coal can energy consumption be sustained through 2050. After that, all every known source of carbon-based fuels will decline rapidly. In 2005, oil prices have been trading at times at their nominal peak from $50 to $60 dollars per barrel. In times past, the Saudi Arabians have increased production to keep oil prices at "reasonable" levels, but it is doubtful that the kingdom has enough excess capacity to do so any more.

Wind energy and hydroelectric power are also ultimately brought about from the power of the sun to warm the atmosphere, causing wind currents, and the transfer of water by evaporation and condensations.

The Earth's tides are caused by the gravitational pull of the moon. These have been used in a small way to generate energy through the use of turbines that run on tidal flows. Likewise, a small amount of energy has been captured from geothermal heat by injecting water into the earth and using the steam that comes out to power turbines.

Atomic energy comes in two forms: fission and fusion. Fission is the result of the destruction high atomic weight molecular elements, like plutonium, as exemplified by the atomic bomb. Atomic energy plants in use today run off the energy obtained by splitting the atom. At one time, it was assumed that atomic energy would be extraordinarily cheap. However, safety concerns and long-term costs such as the disposal and storage of radioactive waste have undermined these rosy assumptions.

Fusion of small atomic weight atoms can also result in the release of extraordinary amounts of energy. This is the basis of the hydrogen bomb. However, fusion has yet to be controlled sufficiently well to allow a safe and efficient fusion reactor for creating electricity.

The world could consume a great deal less energy if we used the energy that we do need more efficiently. I was treated to an illustration of the problem once outside a hotel room in Las Vegas. The temperature inside my room was about 72 °F (22 °C). The temperature in the desert surrounding Las Vegas was about 96 °F (36 °C). But the temperature in the corridor outside my motel room was easily 110 °F (43 °C). To make my motel room cool, it was necessary to pump the excess heat somewhere, which is to say, immediately outside the room. This build-up of heat in the corridor had the effect of instantly increasing the heat gradient against which the air conditioner had to pump. The air conditioners for all of the motel rooms were mounted in the windows and were trying to pump heat into essentially the same space. Add to my motel the hundreds of others like it in Las Vegas and it is easy to see why the city of Las Vegas is almost always hotter than the desert that surrounds it. Not that anybody cares, because almost nobody hangs around outside. Las Vegas proves that humankind could adapt to orbiting space communities as long as we are generously supplied with booze, half-naked ladies and gambling opportunities.

Heat islands exist in all major cities, like New York or Tokyo. We are not just heating up the environment through the carbon dioxide-mediated greenhouse effect; we are doing so directly with the excess heat from air conditioning, manufacturing, and internal combustion machines, not to mention our hot, sulfurous bodies. In 2004, Tokyo – a city set on a windswept island at the edge of the North Pacific – set a record temperature of over 103 °F (39 °C). For the most part, this was not global warming at work. It is Tokyo heating itself. If we could capture and use the energy of waste heat – or better yet, eliminate waste heat – we would be along way down the road to energy independence.

What can nanotech do? Perhaps many things. A workshop held in March of 2004 on the subject sponsored by the Nanoscale Science, Engineering and Technology Subcommittee of the National Science and Technology Council came up with a total of nine basic ways that nanotechnology aid in either improving energy efficiency or providing power. These are listed in Table 21.

Table 21 Energy Grand Challenge research targets.

Scalable methods to split water with sunlight for hydrogen production
Highly selective catalysts for clean and energy-efficient manufacturing
Harvesting of solar energy with 20 % power efficiency and 100 times lower cost
Solid-state lighting at 50 % of the present power consumption
Super-strong, light-weight materials to improve efficiency of cars, airplanes, etc.
Reversible hydrogen storage materials operating at ambient temperatures
Power transmission lines capable of 1 gigawatt transmission
Low-cost fuel cells, batteries, thermoelectrics, and ultra-capacitors built from nanostructured materials
Materials synthesis and energy harvesting based on the efficient and selective mechanisms of biology

Source: Nanoscience Research for Energy Needs: Report of the Nanotechnology Initiative Grand Challenge Workshop, March 16–18, 2004.

At present, one of the major uses, by dollar volume, of nanotechnology is the employment of nanoscale catalysts in petroleum distillation. Materials that have little or no catalytic activity bulk form can deliver exceptional catalytic behavior in nanoscale form. In part, this is due to surface effects.

In February of 2005, a transportation company called The Stagecoach Group announced that it would begin using a nanoparticle-based catalyst in its 7000 vehicle fleet. Called Envirox, this cerium oxide-containing nanoparticle is manufactured by Cerulean International, a subsidiary of Oxonica Ltd. It has been recognized for some time that cerium oxide could give a cleaner-burning fuel, but until nanoparticles could be manufactured, the catalyst simply settled out to the bottom of the gas tank. Nanoparticles are small enough to stay in solution. The catalyst increases fuel efficiency by 5 % – not an answer to the world's energy problems by any means, but at least an incremental improvement. Not a great leap forward perhaps, but a technological advance that is ready for deployment.

Table 22 indicates other areas in which nanotech might be employed, and some of the proponents of each technology. Taken together, these innovations, if successful, could do much to make energy independence a reality.

Table 22 How can nanotech help to produce or save energy?

Concept	Method	Proponent(s)
More efficient combustion of carbon fuels	Envirox – cerium oxide nanoparticulate catalyst	Oxonica
Use "quantum pipes" for energy transmission through power lines to reduce energy loss.	Create ultra-efficient power lines using carbon-nanotubes	Richard Smalley, Rice University
LEDS to replace incandescent light	Polymer LEDS Quantum dot LED	Cambridge Display Cree Research
More efficient solar panels	Thin-layer polymers to translate light into electricity	Cambridge Display Technologies
Nanotech-enabled fuel cells	"Nanocubes" or other nano-materials to store hydrogen	BASF, Nanosys
Photo-induced release of hydrogen from water	TandemCell employing metal oxide conversion of sunlight to electricity; electricity to split water	Altair Nanotechnologies
	Supermolecular complex combining photosynthesis and electron transport	Researchers at Virginia Polytechnic and Virginia State Universities
	Nanotubes as solar collectors	Penn State University
Green energy	Use of microbes to make energy compounds	J. Craig Venter Institute, Solazyme

Nobel laureate Richard Smalley, co-discoverer of the buckyball, has proposed an elaborate system to solve the world energy crisis that would require something on the order of an Apollo project to put in place. Students need to be inspired to enter the sciences, as they were after the launch of Sputnik was revealed. "Be a scientist," says Smalley, "Save the world." He has proposed a tax on gasoline of 5 cents per gallon to finance his program.

Smalley proposes the transfer of carbon-based forms of energy – coal, oil and gas – around the globe be largely eliminated. Instead, energy could be produced locally and transmission of energy would be in the form of electricity through large power grids. Of course, we already have power grids all over the world, with varying degrees of reliability. One of the problems with "wheeling" power from one region to another has always been the dissipation of energy through heat loss in the wires. One kilowatt of power bought in St. Louis is not really a kilowatt anymore by the time it arrives in New York City. Smalley's solution is to rewire the electrical grids with cables made from carbon nanotubes. These, he points out, have the conductivity of copper with one-sixth the weight, the thermal conductivity of diamond, and are theoretically the strongest fibers it is possible to make. "Quantum wire" made from carbon nanotubes, says Smalley, would have negligible "eddy current" loss and should allow a vast improvement in the efficiency of energy transmission.

Another aspect of Smalley's energy revolution would be the development through nanotech of ways to generate and store energy locally. Deregulation of the energy markets in many parts of the U.S. already allows some consumers to generate electricity through wind power, biomass conversion, or cogeneration and to sell it back to the grid. A sticking point here is that there is currently no good way to store large volumes of energy. If there were, it would be possible to generate power during periods of low demand and to sell it during periods of high demand, thus optimizing the system.

Electricity use accounts for about one-third of total energy consumption in the U.S. and presumably, in the rest of the industrialized world. About 20 % of all electricity consumed goes for lighting. However, today's lighting is remarkably inefficient. Incandescent lights have a luminous efficiency of 15 lumens per Watt, and fluorescent lights a luminous efficiency of 80 lumens per Watt. Only about 5 % of electrical energy used in incandescent bulbs is converted into light. Florescent lights are better (if you can stand the flickering and the hum), but they still have an efficiency of 25 %. These are mature technologies that have been pushed about as far as they can go. It is anticipated that the use of LEDS for general lighting could increase overall efficiency by 50 %.

Monochrome LEDs are achieving energy efficiencies as high as 50 % in the red and on the order of 20–25 % in the blue. LEDs are already 10 times more energy-efficient than their incandescent counterparts, and they have already replaced over one-third of the traffic lights in the U.S., saving about $1000 per intersection per year in electricity. Achieving acceptable white light for general illumination requires an efficient blue LED, which does not yet exist.

Thin-film polymer-based LEDS are still too energy-inefficient to be competitive with white light, but the technology is only a few years old, and it is expected that further refinements could push it past phosphorescent lights. As polymer LEDS come in colors, three colors that add up to white light must be used. Quantum dot-based LEDs are another possibility. Cree Research has produced a white light LED using nanocrystalline quantum dots as phosphors, although quantum dots are still very expensive to manufacture.

Nanotech would allow new ways of using the sunlight to generate power, including improved photovoltaics (solar panels). We have already seen, for instance that polymers in nanoscale thin layers can be used to generate electricity in response to light; the reverse action of the polymer LEDs invented by Cambridge Display Technologies (see Chapter 7).

Photoelectric power can also used to liberate hydrogen from water; the hydrogen could then be used in fuel cells. Hydrogen, as a fuel, suffers from two main problems. First, it is very explosive; and second it is the lightest element and exists as a gas at normal temperatures. In order to store hydrogen, it has traditionally been necessary to either cool it to extremely low temperatures or to force it at high pressure into reinforced containers. An alternative involves absorbing hydrogen onto nanoscale surfaces. Carbon nanotubes have been proposed as "molecular sponges" to hold hydrogen under relatively low pressure to allow the manufacture of hydrogen-burning vehicles. Researchers from BASF have developed "nanocubes", which are crystalline structures with a very high surface area. These, it is believed, could be used to store hydrogen for use in fuel cells used to power consumer devices, such as cell phones and laptop computers. The corners of nanocubes are composed of zinc atoms linked by organic acid molecules in a kind of grid or lattice. Because the hydrogen is adsorbed to a surface, it actually occupies much less space than it would as a gas, and requires less pressure to keep it contained.

Nanosys, a California-based nanotech firm, is also creating nanomaterial-based fuel cells in collaboration with Japanese consumer electronics company, Sharp Corporation. Nanosys has also received a $14 million contract from DARPA to create flexible, low-cost solar cells.

A U.S. government program, "the Freedom Car" has been initiated to deliver a prototype hydrogen fuel cell-powered car in five years. If an economical way can be found to deliver and store hydrogen, such a car is much to be desired as the combustion of hydrogen yields water, rather than hydrocarbons. There has been much speculation about a "hydrogen economy" that will supersede the carbon fuels era. But first, of course, a way must be found to generate hydrogen without using carbon fuels as the ultimate energy source.

Altair Nanotechnologies and Hydrogen Solar are collaborating to make a photon-powered hydrogen generation system using nanomaterials. They are collaborating on a product called a Tandem Cell, that utilizes thin-film metal oxides that use the energy provided by ultraviolet and blue light to generate electron-hole pairs. Longer wavelength light passes into the second part of the device called a Graetzel cell, to create a voltage potential. The Graetzel cell employs a thin film of titanium dioxide with a dye superimposed. Together, the twin cells create a kind of battery which is continuously recharged by solar energy. The current generated is used to split water electrolytically into hydrogen and oxygen. The companies estimate that a garage roof could provide sufficient surface to create enough fuel to drive a car about 11 000 miles a year.

The first law of thermodynamics prevents you from obtaining more energy from hydrogen or methanol than you invest in order to obtain them, but if the energy source you invest comes from the sun, these are potentially very cheap sources of energy. Researchers at Virginia Polytechnic and Virginia State universities are working on a supramolecular complex that uses energy from the light-catalyzed extraction of hydrogen from water.

Penn State University researchers have utilized titanium nanotubes to collect ultraviolet light and then use its energy to extract the hydrogen in water. The efficiency of UV energy capture was 97 %, a very high value, but the efficiency of hydrogen extraction was only 6.8 %. Plus, the process wastes the 95 % of sunlight that is not in the ultraviolet range.

Green Energy

Back in the 1970s, a University of San Diego biology professor named Gordon Sato (whose efforts vastly improved the practice of mammalian cell culture) came up with an idea to solve what was then believed to be an urgent energy crisis – called by President Jimmy Carter "the moral equivalent of war." Sato proposed that an unused portion of the California desert should be flooded to a depth of 6 inches (15 cm) or so and then seeded with photosynthetic algae that had the ability to synthesize hydrocarbons, using carbon dioxide from the atmosphere as a carbon source. The algae could simply be harvested, and the hydrocarbons extracted.

Nobel laureate Hamilton Smith and Craig Venter, who were largely responsible for directing the sequencing of the human genome at Celera, have come up with a new twist on Sato's idea. Smith is head of the Synthetic Biology Group at the J. Craig Venter Institute. They intend to co-opt a minimal genome from a type of mycoplasma, a small bacterium. The idea is to create a synthetic genome, which they will introduce into an artificial cell. The organism would only have only the genes required to maintain and reproduce itself. They will then adapt this microbe for other purposes. The Synthetic Biology Group is engineering new pathways that could lead to new methods of carbon sequestration; for example, taking carbon dioxide out of the air. Their stated goal is to create a bio-derived alternative energy source. Cognizant of potential ecological damage, the team is developing only synthetic organisms that completely lack the ability to survive outside the laboratory.

Solazyme, an entrepreneurial company backed by Harris & Harris, among others, is investigating ways of converting carbon dioxide into carbon fuel compounds through the use of photosynthetic organisms. So far, they have released few details about their research program.

Shuguang Zhang at MIT and his research collaborators have integrated a protein complex derived from spinach chloroplasts with organic semiconductors to make a solar cell that could be combined with solid-state electronics.

Chloroplasts are the organelles in plants cells that are packed with chlorophyll – the molecule that makes plants green and allows them to carry out photosynthesis. Zhang's team has managed to artificially stabilize the protein complex at the heart of their system – comprised of 14 protein subunits and hundreds of chlorophyll molecules – using synthetic peptides to bind small amounts of water to it, within a sealed unit. Plants use photons, to excite coupled pairs of electrons within chlorophyll, causing an electron to transfer to a nearby receptor molecule. Energy thus extracted from the sun allows plants to convert carbon dioxide into sugar molecules. The device developed by Zhang uses the same process to feed electrons into organic semiconductors. Right now, Zhang's green solar cell is more of a research project than a practical device because of its limited stability and low efficiency.

The Space Elevator

About 4000 years ago, a man named Jacob (Genesis 28:12) had a dream about a ladder which stretched from the Earth to the heavens. The angels of God were ascending and descending on this ladder. By 1978, as technology had improved, the updated version of the ladder became an elevator in Arthur C. Clarke's novel the *Fountains of Paradise* [1]. Imagine climbing into an elevator at an embarkation point somewhere in the Pacific Ocean and climbing into deep space. The elevator would ride up on a slender ribbon hanging in space, in delicate balance between the force of gravity and the centripetal acceleration of the spinning Earth. Like a giant spider web, the ribbon would stretch from a large ship anchored near the equator in the Pacific Ocean up 62000 miles through the atmosphere to a platform orbiting in space (Fig. 45). The center of mass of the whole complex would be in geosynchronous orbit (GEO).

Actually, the space elevator concept springs not solely from Clarke's very fertile imagination, but was first described in a journal article by Russian engineers some fifty years ago. Yuri Artsutanov gave the first detailed explanation of the idea in 1960 [2]. Another Russian, F. Tsander described an Earth to Moon cable tether as far back as 1910. However, it was not until the mid-1990s that the NASA Institute for Advanced Concepts began to take the idea seriously and commissioned aeronautical engineer Brad Edwards to carry out a feasibility study. Today, Edwards is head of a private company called Carbon Designs that is dedicated to making the space elevator a reality. A former associate, Michael Laine, heads LiftPort Group, a kind of miniature technology conglomerate that is also focused on the space elevator. There is an international Space Elevator Conference going into its fourth year.

The discovery of the carbon nanotube is what brought the space elevator out of the realm of science fiction. So far, carbon nanotube fibers are the only material conceivably strong enough to form the elevator ribbon. The ribbon has to deal with enormous forces – gravity in one direction, and the centripetal force caused by the rotation of the Earth. The vacuum of space poses other problems: localized heat caused by the action of the lifters is difficult to dissipate, for instance. Then there is the problem of wind resistance in the atmosphere, lightning strikes and, potentially, collisions with aircraft. In space, the ribbon could be struck by meteorites and would be constantly bombarded with electromagnetic radiation.

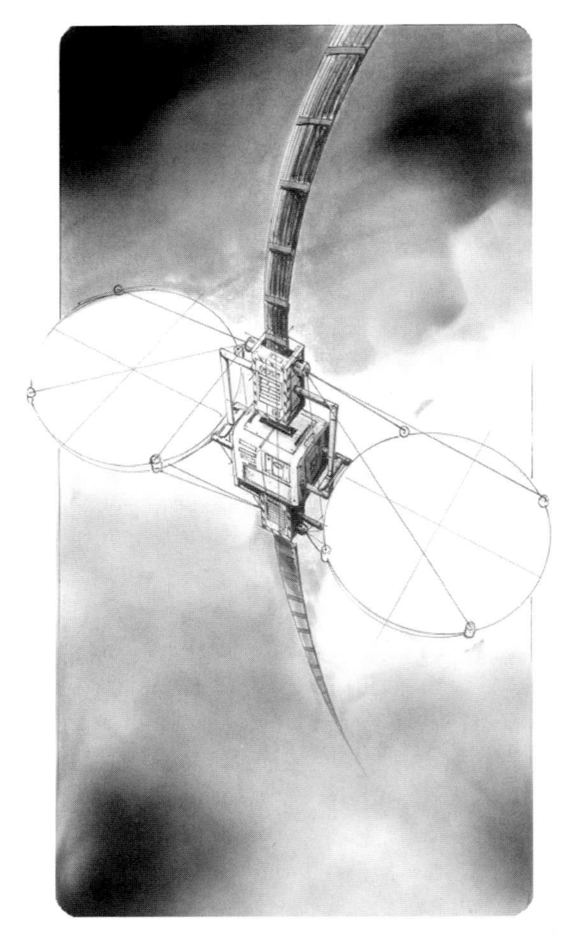

Figure 45 Lifter ascending though the clouds on space-elevator ribbon. Drawing by Nyein Aung. Image reproduced courtesy of LiftPort Group.

Michael Laine, CEO of LiftPort, envisions that an initial deployer spacecraft would be placed in low Earth orbit via rocket launches and assembled there, possibly with the assistance of a space shuttle crew. The deployer would then use an electric propulsion system to climb gradually to a high geosynchronous orbit, about 22 300 miles above the Earth. Once there, it would begin playing out a flat ribbon composed of carbon nanotubes, about 3 feet (1 meter) wide but paper-thin, down into the atmosphere. The deployer would gradually climb to about 62 000 miles, the full extent of the ribbon. The 80-ton deployer craft would then permanently serve as a counterweight at the end of the ribbon. In theory, the ribbon could be shortened by introducing a heavier counterweight, but there is an advantage also in having a long ribbon, in that it allows greater access to deep space.

The first ascending missions would add more ribbons to increase the strength of the cable. One could imagine eventually a kind of space station built around the GEO point, a convenient stop-off point for servicing satellites and other orbiting craft, and perhaps a loading zone for climbing on the elevator half-way up.

The whole complex, an 800-ton mass of cable and equipment, would be exquisitely balanced by opposing forces. The force of gravity at the lower end of the nanotube cable and centripetal acceleration at the farther end (imagine swinging a weight on the end of a rope) keep the nanotube cable under tension and hopefully stable in its stationary orbit over Earth.

Simply by stepping off the elevator one could assume a specific orbit at various distances from the Earth. Traveling to the end of the ladder would allow one to escape the gravity well of Earth altogether. As you near the end of the elevator cable, centripetal force overcomes gravity, and your vehicle accelerates to extraordinary speed, over 20 000 miles per hour, allowing you to sail off into space after you disengage. In essence, you would be using the force from the Earth's rotation to whip your vehicle off into the void.

The "lifters" that would ascend the elevator cable would be essentially big gondola cars, that roll up the cable on wheels. Their propulsion system would involve an Earth-based free electron laser, in Laine's version, the beam of which would be converted to electricity to power the vehicle.

Why do we need a space elevator when we have a space shuttle already? Well, cost is an issue. To put 1 lb (0.5 kg) of material into a low Earth orbit costs about $100 000. The space elevator could achieve FedEx price levels, delivering the same package for only $100, cheaper by a factor of 1000, according to its proponents.

Then there is the issue of safety: if the space shuttle doesn't blow up on the way up, the astronauts' next most dangerous problem is re-entry. The elevator can also go down as well as up, bringing you back to Earth at a gentle 10 miles an hour if you choose to, according Brad Edwards.

There are, of course, substantial obstacles in implementing the Space Elevator design, not the least of which is manufacturing a sufficient quantity of nanotubes and spinning them into a cable. Michael Laine estimates that the ribbon, thin as it is, would nonetheless contain 600 tons of carbon nanotubes, which LiftPort plans to manufacture on its own. To do this, the company has to mass produce large quantities of nanotubes with a high yield of quality-controlled product. These are problems not yet solved by a myriad of nanotube companies which, to date, can provide only kilogram quantities of nanotubes of varying quality.

It is not clear yet whether the space ribbon will be constructed from single-walled or multi-walled nanotubes. The argument for the former is that there would be greater quality control. Multi-walled nanotubes could, in theory, provide greater strength, provided that they could be modified such that the tubes would not slide past each other. At this point in time, it is easier and cheaper to produce multi-walled tubes by vapor deposition than it is to produce single-walled tubes, which are generally made by laser ablation in a tube furnace.

A group called Elevator 2010 (www.elevator2010.org), associated with the Spaceward Foundation, holds an annual contest with a $50 000 prize to the person or group that can manufacture the strongest space elevator ribbon (tether) or build the best lifter (climber). The lifters only need to climb about 200 ft (60 meters), however, not 62 000 miles. The contest is held in the San Francisco Bay area, and is funded with help from NASA, according to Ben Shelef, an engineer from Gizmonics, and a spokesman for the conference. Shelef"s company is an enthusiastic supporter of the space elevator concept and has already contributed some prototype designs for the project. These can be viewed on the web at www.gizmonicsinc.com/elevator.

Before carbon nanotube composites came along, the strongest fiber that could be made was Kevlar, with a tensile strength 3.5 gigaPascals (gPa). Simulations of carbon nanotubes indicate a tensile strength of 130 to 300 gPa. So nanotube fiber composites, possibly including epoxy, are expected eventually to be the material that makes the ribbon to the heaven's possible. The exact nature of the weave is still up for debate. Minimizing damage from meteorites is a serious consideration. The ribbon envisioned in Brad Edwards' original report to NASA was to be 2 inches (5 cm) wide at the base, tapering to 5 inches (11.5 cm) at the geosynchronous orbit point, but only 1 μm (one-millionth of a meter) thick.

The laser-powered lifter technology is also largely theoretical at this point. However, a LiftPort demo model robotic lifter was able to climb a cable suspended from MIT's Green Building, a distance of about 290 feet (90 meters). The first 10 % of the elevator-ride into space, when the lifters are fighting the Earth's gravity, would use almost all the power. After you pass the midpoint, the ride is energetically free, as the rotation of the Earth provides the power to whip the lifters higher up the ladder.

Then there is the problem of raising the money; both Laine and Edwards estimates that the first such elevator with a 20-ton capacity will cost in the neighborhood of $10 billion – not outrageous by the standards of the space transportation industry, but a considerable obstacle for a small company with no revenues. In Edwards' view, public funding of the project is unlikely at present. Venture capitalists are generally not enthusiastic about fifteen-year projects with an unpredictable rate of return. Laine's response has been to divide his project into chunks with technology that can be spun off for other applications. So LiftPort Group consists of LiftPort Carbon, LiftPort Robotics, LiftPort Media and LiftPort Finance, in addition to LiftPort Inc., which has overall responsibility for the Space Elevator. LiftPort Carbon is charged with level production of high-quality nanotubes, for which there is already a market. LiftPort Robotics would be responsible for the development of lifter technology. LiftPort Media is a publishing company that will publish books and documentaries about the Space Elevator project, along with calendars and posters. LiftPort Finance is being formed to encourage public investment in space enabling enterprises. They plan to offer the LiftPort Space Technology Mutual Fund, a LiftPort Finance Venture Capital Fund focused on start-up aerospace companies, and as well as limited partnerships. Laine hopes to tap into rich folks of the Bill Gates or Paul Allen variety – people with so much money that they wouldn't mind using some of it to support worthy, if somewhat risky, ventures.

Just like the Apollo project, the Space Elevator project is likely to generate substantial returns here on Earth. The nanocomposite ribbon alone would have innumerable uses in constructing suspension bridges, monorails, unbreakable high-tension lines, etc. Nanotube cables could even be used to suspend large buildings from tree-house-like platforms, saving space on the surface of the planet for important things, such as plants and animals.

The need for tons of nanotubes is what makes the Space Elevator a nanotech project: LiftPort is already building a full-scale commercial nanotube factory in Millville, NJ in June, 2005. For its part, Carbon Designs has licensed nanotube technology from Los Alamos Laboratory.

Besides nanotubes, other aspects of the technology required for the space elevator are just as daunting, including the exact design of the ribbon, the power source for the lifters, and the problem of scale-up. Another serious problem may be political. To whom do you apply for permission to build a 62 000-mile elevator into the sky? Surely, some earthbound governmental body will want an environmental impact statement, at the least. If the thing collapses, the Earth could end up spooling a nanotube ribbon wound twice around its equator with an 80-ton weight flopping around at its end. There would have to be substantial guarantees assured to those possibly affected. Through public relations programs, LiftPort hopes to create an enthusiastic consensus and hopefully some government help to bring the program to fruition.

Michael Laine and the LiftPort group remain optimistic. As this is written, they are one year into a fifteen-year plan – deployment has already been scheduled for April 12, 2018. Laine, like all good entrepreneurs, suffers from pathological optimism. *Nanotech Pioneers* is willing to be generous. We will consider our challenge met if LiftPort or any other organization has lifters climbing into space by the year 2025.

Building a Quantum Computer

If there is an ultimate killer application for nanotechnology it is probably quantum computing. Imagine computers millions of times more powerful and thousands of times smaller than the computers we have today. Computer scientists have been dreaming about the quantum computer since the 1980s, and prototypes that can do a few calculations before they fall apart have actually been built. But to build one that is useful will undoubtedly require a heavy dose of nano.

Incidentally, DARPA was reportedly thinking about a funding a "moon-shot"-type program to create a quantum computer in 2004, but backed off when researchers warned that the technology was not yet ready. We at *Nanotech Pioneers* are prepared to take that leap-of-faith and propose a quantum computer Grand Challenge (as long as somebody else will provide the funding and do the work).

Computers, of course, have been scaling down for decades. The first computers had about 18000 vacuum tubes, 500 miles of wiring, and weighed 30 tons. The first computer that I ever saw (in 1969) occupied a floor of the mathematics building at the University of California, San Diego. It was quite a sight; row after row of circuit boards, switches and breakers, wires draped everywhere, with big ducts intersecting it at intervals. The whole floor below it was taken up with air conditioning to keep the thing cool. If you wanted the Behemoth to do anything, you had to feed it punch cards. As undergraduates had absolutely the last priority on this shared facility, I invariably had to show up at midnight for my appointment with this icon of high technology. Inevitably, the machine would send me back an error message informing that I'd punched a card wrong; then I would have to fix it and come back the next night. I hated the thing.

This book is written on a Dell computer, the latest in a series of personal computers I have owned going back to my first, beloved Apple II. The latter had no hard-drive, and if I wanted a color monitor, I had to hook it up to my TV set. Still, it was nearly as powerful and certainly more useful than the Behemoth in the UCSD maths building.

Still, every computer from the world's first electronic digital computer, ENIAC, built in 1945, on down to the present is more or less a faithful representation of the "Analytical Engine" that Charles Babbage was unable to complete before his death in 1871. Babbage's machine would have been mechanical rather than electronic, and so, one assumes, quite a bit slower. Computer engineer Danny Hillis actually built a mechanical computer out of Tinkertoys when he was an undergraduate in 1975. K. Eric Drexler designed a nanoscale mechanical computer when he was a student at MIT [3], although, like all of Drexler's work, it never got off the drawing table. Mechanical or electronic, it doesn't really matter. The task of a computer is to manipulate binary bits (1s and 0s) into a useful computational result.

Quantum computers are different – they use quantum bits, or qubits. A qubit can be represent a 1, 0, or some combination of both. The following explanation comes from Hans Moravec [4]:

"Like a conventional computer a quantum computer consists of a number of memory cells whose contents are modified in a sequence of logical transformations. Unlike a conventional computer, whose memory cells are either 1 or 0, each cell in a quantum computer is started in a quantum superposition of both 1 or 0. The whole machine is a superposition of all possible combinations of memory states. As the computation proceeds, each component of the superposition individually undergoes the logical operations. It is as if an exponential number of computers each starting with a different pattern in memory, were working on the problem simultaneously. When the computation is finished, the memory cells are examined, and an answer emerges from the wavelike interference of all possibilities. The trick is to devise the computation so that the desired answers reinforce, while the others cancel."

If that explanation seems confusing, it is because it is. "If you think you understand quantum mechanics," said Richard Feynman, "then you don't understand quantum mechanics."

It is hard to be comfortable with the idea that a qubit can be both one and zero, or anything in between all at the same time. "Do not [ask], if you can possibly avoid it, but how can it be like that?" warned Richard Feynman, "because you will go down the drain into a blind alley from which no one has yet escaped. Nobody knows how it can be like that."

When the memory elements are interrogated at the end of the process, the quantum uncertainty collapses, revealing the answer. In effect, each qubit essentially acts as a parallel processor that can do many computations at once. A quantum computer with only 30 qubits, in theory, could equal the processing power of a conventional computer that ran at 10 teraflops (trillions of floating-point operations per second). That is over 1000 times as fast as today's desk-top computers. A quantum computer with 300 qubits could perform operations on 2^{300} machine states at once and would be faster than any conventional supercomputer imaginable.

Richard Feynman, naturally, was among the first to appreciate the potential of the quantum computer. In a paper in 1982, he showed that a quantum computer could be used to perform computations, and further that it could be used as a simulator for quantum physics. David Deutch followed up with a paper in 1985 showing that quantum computer could model any physical process and serve as a general-purpose computer.

Incidentally, Roger Penrose, in his book, *The Emperor's New Mind* [5] suggests that human thought employs quantum mechanics. He proposed that the physiological process underlying a given thought may initially involve a number of superposed quantum states, each of which performs a kind of calculation. At some point, the superposed states collapse into a single state, causing measurable and possibly distributed changes in the neural structure of the brain. The Emperor's "new clothes" in the ancient fable, of course, were not there. Penrose's title is a dismissal of the abilities of computers to act in a brain-like fashion, because they don't have this supposed quantum capability. Yet.

How Do You Build a Quantum Computer?

All theory aside, how do you actually build a quantum computer? So far, only computers of a few qubits have been constructed, and these are highly unstable. Some researchers believe that quantum computers will never be more than a curiosity for physicists to play with, while others seem to regard their development as a commercially viable product as being inevitable. The biggest problem is one called "decoherence." Quantum state superposition is hard to maintain for any length of time (as in nanoseconds) because the particle (an atom or electron or even a photon) is constantly interacting with its environment, leading to a collapse into a determinate state, interrupting the calculation in progress.

Electrons are particularly attractive for representing the qubit because they have the quantized property of spin (this is discussed in Chapter 7). The two states are referred to as "up" or "down." Imagine that you are the nucleus and that the electron is spinning from bottom to top as it simultaneously orbits around you. That would be the "up" spin. Now, imagine it spinning the other way, from top to bottom – that's the "down" spin. Now imagine it spinning horizontally parallel to its orbit – that's the indeterminate spin, the superposition of both states. Actually, these visual images probably have no correlation with reality, but they help us to imagine what is happening.

Electronic methods that have already been developed can detect the flavor of spin that an individual electron has. Electrons for use in quantum computing could be trapped in quantum dots or quantum wells (as discussed in Chapter 4). Physicist Albert Chang of Duke University and his colleagues have made qubits from quantum dots by placing dots containing "puddles" of electrons with the same net spin value adjacent to one another and connecting them with tiny wires. The quantum dots are about 200 nm in diameter and about the same distance apart, so this was quite a feat of fabrication. By controlling the charge that passes between the dots, the team was able put them into a "coherent" state of indeterminate spin – spinning up and down simultaneously, just the property you need for good qubits.

Quantum dots are perhaps the best current option for creating quantum computers, because it is already possible to fabricate large arrays of quantum dots. A team at the University of Wisconsin, led by Mark Eriksson of the Physics Department, has determined by modeling that a million-quantum-dot computer (an 1024×1024 array) could be built using current technology. They warn that the device would operate only in the megahertz range, instead of gigahertz, due technical device constraints. However, because each qubit is essentially a parallel processor, it would still be blazing compared to your average laptop. Though a long way from a commercial model, the device could serve as a prototype operable quantum computer and perhaps perform useful calculations that require a high degree of parallelism and are difficult with conventional computers. The team is now hard at work building a prototype.

Another method that is being pursued for quantum computing involves ions (beryllium and calcium have been used) held in an electromagnetic trap. In this situation, their quantum states can be manipulated with laser beams and their physical movement can be controlled with electrodes. One version of such a device, developed at the National Institute of Standards and Technology, employs three ions per qubit. One is the actual workhorse qubit and the two remaining are there for error correction in case something happens to the first one. Like electrons, atoms have an angular momentum or "spin" that can be either up or down, and this is the property used for computations. Certain operations allow the researchers to link the spins of the three ions through quantum "entanglement", meaning that an operation on one will automatically be reflected in the other two.

The quantum entanglement effect, an action at distance with no obvious mediator, is so weird that Einstein himself called it "spooky" and no one has yet proposed a mechanism that can be understood by mere mortals. Suffice to say that it works. The object of the entanglement, in this case, is error correction. At the time of measurement, the primary qubit is disentangled from its partners. The states of each ion are measured. If the primary ion differs from its two partners, it is a sign that an error has occurred during computation, and the error can be detected and the system reset. Entanglement of as many as five particles has been demonstrated, which allows even more fault-proof error containment, presumably.

Entanglement of quantum states has actually been demonstrated between entrapped ions and photons. This, in theory, would allow the "teleportation" of quantum information over optical networks.

Objects as large as buckyballs (60 carbon atoms) have been shown to exhibit quantum effects. One candidate that has been suggested for the qubit is a carbon nanotube. Consider what happens when you push on the ends of a plastic ruler. The ruler, of course, bends: but which way, to the right or to the left, away from you or towards you? The answer is hard to determine without doing the experiment. When you push on the ends of a carbon nanotube (a perfectly symmetrical object), because of quantum effects it should occupy all possible states simultaneously. In theory, it would bend both left and right and back and forth all at the same time; its position would be a quantum blur until some observational process was brought into play.

Nanoscale devices are likely to be required to build a useful quantum computer, in that they are small enough to exhibit or sense quantum effects, but still large enough to be accessible for human control. In December 2004, the Semiconductor Industry Association (SIA) launched the Nanoelectronics Research Initiative designed to link industry, academia and the federal government into a mission-oriented effort to keep Moore's law going on and on into the future. By the year 2015, it is estimated, a transistor will have to shrink to a single-electron device for the incredible shrinking chip to continue. How long can it go on? "We need 300 Moore's law doublings, or 600 years at one doubling every two years," calculates MIT's Seth Lloyd, "before all the available energy in the universe is taken up in computing [6]."

If quantum computing takes off, will the SIA be unpleasantly surprised to discover that its semiconductor fabs have suddenly been made obsolete prematurely? Probably not, at least in the near future. The first quantum computers will be special- purpose devices designed for heavy-duty number-crunching, such as is needed for weather prediction or cryptography. Indeed, the general availability of quantum computers would render present schemes of cryptography, which involve the factoring of incredibly large numbers, useless.

"A quantum computer, if it works, in principle would obsolesce all the encryption in the world today," said former U.S. Speaker of the House, Newt Gingrich. "That's a pretty big deal. If I said to you that the U.S. could have the capacity to read any encryption in the world and do so in about 3 seconds or less, what's that worth? Well, it's enormous. The challenge for us is: What if the Chinese or the Indians get quantum computing before we do? That would be sobering..." Governments around the world, as you can imagine, are paying heavy attention to the development of quantum computing.

A quantum computer is a completely different beast to a conventional digital computer. It will take some time before languages and protocols are available that would allow whiz-kid programmers to use quantum computers to create the most awesome video/audio sense-surround games you have ever imagined. But once the computers are available, believe me, it will happen.

References

1 Clark, Arthur C. *Fountains of Paradise*, Harcourt, NY **1979**.
2 Artsutanov, Y. V Kosmos na Elektrovoze, Komsomolskaya Pravda, **1960**.
3 Drexler, K.E. Rod Logic and Thermal Noise in the Mechanical Nanocomputer". In: Carter, F.L., Siatkowski, R.E., Wohltjen, H. (eds.). *Molecular Electronic Devices*. Elsevier Science Publishers, B.V., North-Holland, **1988**, pp. 39–56.
4 Moravec, H. *Robot*. Oxford University Press, **1999**.
5 Penrose, R. *The Emperor's New Mind*. Oxford University Press, **1990**.
6 Quoted in: *God is the Machine*. Kelly, K. *Wired* 10: 12 **2005**.

Chapter 11
Fear of Nano: Dangers and Ethical Challenges

K. Eric Drexler's warnings about the possibility of out-of-control, replicating nanobots set the stage for public concern over nanotechnology even before there was much real technology to worry about. Drexler, himself, has recently backed off of his original version of molecular nanotechnology, which emphasized the necessity of self-replication, but the damage has been done. Science fiction authors have had a field day, and warnings about the dangers of nanotechnology have come from not only radical environmental groups but also techno-gurus such as Bill Joy and even Ray Kurzweil.

Recently, a group called Topless Humans for Natural Genetics, or Thong, disrupted a Chicago Nanotechnology Conference with a quick strip-tease revealing Richard Feynman's prophetic words, "There's plenty of room at the bottom" painted on their derrières. The same group later did another strip act in the front windows of an Eddy Bauer store to protest about that establishment's stain-resistant NanoTex pants. "Expose the truth about nanotech," read one of the slogans this time. What truth are they so concerned with? Are stain-resistant pants a danger to mankind?

Public concern has been inflamed by popular entertainment. For instance, replicating nanobots formed the plot of a *Star Trek: Next Generation* episode when Wesley became distracted and left his science experiment unattended for too long. The *Enterprise* was in peril as the nanobots started disassembling everything in sight, beginning with the ship's electronics. Fortunately, by the end of the hour, the nanobots had developed intelligence and a communication system and Data was able to talk them into disembarking on an uninhabited planet. The most extreme disaster nanobots scenario, though, was laid out by John Robert Marlow in his sci-fi novel *Nano* (Tom Dougherty and Associates, New York, NY, 2004). Voracious nanobots eat up a good part of the San Francisco Bay area, and they are barely stopped from spreading worldwide by the scientist who invented them. He then declares himself dictator of the world, because, of course, mankind is too greedy and evil to deal with the power inherent in nanotechnology. Any complex technology, he claims, requires totalitarian institutions to manage it (sorry to have given away the plot, but it's a pretty lame novel anyway). But the question it raises cannot be easily dismissed: Are draconian regulations required to keep technology, particularly nanotechnology, from causing harm?

The Nanotech Pioneers. Steven A. Edwards

ISBN: 3-527-31290-0

The dangers of the uncontrolled distribution of technology were also addressed by Bill Joy [1]:

"The 21st-century technologies – genetics, nanotechnology, and robotics – are so powerful that they can spawn whole new classes of accidents and abuses. Most dangerously, for the first time, these accidents and abuses are widely within the reach of individuals or small groups. They will not require large facilities or rare raw materials. Knowledge alone will enable the use of them."

There is already considerable concern over the potential for terrorist groups to acquire biological weapons. A single smallpox-infected terrorist armed with a tourist visa and a Eurail pass could start a pandemic in Europe, for instance.

Etc (pronounced *et cetera*), a technology watchdog type of organization that was prominent in the fight against genetically engineered organisms, has now put its sight on nanotech. They have published an 84-page report titled *The Big Down – Atomtech: Technologies Converging at the Nanoscale* [2]. Not surprisingly, they find nanotech alarming. For example, this excerpt:

"The hype surrounding nano-scale technologies today is eerily reminiscent of the early promises of biotech. This time we're told that nano will eradicate poverty by providing material goods (pollution free!) to all the world's people, cure disease, reverse global warming, extend life spans and solve the energy crisis. Atomtech's (Etc speak for nanotech) present and future applications are potentially beneficial and socially appealing. But even Atomtech's biggest boosters warn that small wonders can mean colossal woes. Atomtech's unknowns – ranging from the health and environmental risks of nanoparticle contamination to Gray Goo and cyborgs, to the amplification of weapons of mass destruction – pose incalculable risks."

Etc offers as a guiding concept its Precautionary Principle which says that "governments have a responsibility to take preventive action to avoid harm to human health or the environment, even before scientific certainty of the harm has been established. Under the Precautionary Principle it is the proponent of a new technology, rather than the public, that bears the burden of proof." Proving a negative, as any scientist or lawyer will tell you, is an almost impossible task. That's why, in law, there is the presumption of innocence. How can you prove that any technology is totally safe in advance of its use?

Etc wants an immediate moratorium on commercial production of new nanomaterials. Molecular manufacturing, says the group, poses "enormous environmental and social risks and must not proceed – even in the laboratory – in the absence of broad societal understanding and assessment."

Etc is not alone in its concerns. A report from Abdallah Daar and Peter Singer, from the University of Toronto Joint Centre for Bioethics, has called for a general moratorium on nanomaterial deployment, on the grounds that the ethical, environmental, economic, legal and social implications of nanotechnology have not yet been taken seriously or pursued on a large enough scale.

Etc and groups like them cannot be ignored, if only because they have demonstrated the power to sway public opinion. Under its previous incarnation as RAFI, Etc group members were given credit for putting a halt to Monsanto's so-called "Terminator Technology," a genetic engineering method to protect agritech patent rights by making second-generation seeds sterile. Etc and similar organizations have recently won a powerful convert in Prince Charles of England. That a member of a vestigial monarchy – a living anachronism – inveighs against the horrors of modern technology is not without irony. But he has a certain amount of influence. Public and political opinion can matter a great deal, as agritech and stem cell companies have learned by bitter experience. From the standpoint of companies trying to commercialize nanotechnologies, risks perceived by the public, whether realistic or not, can be just as destructive as actual, quantifiable risk due to the nature of the technology.

Is nanotech dangerous? And if so, is government monopolization of its use required? In effect, this is already a moot point as the genie is out of the bottle; nanotech is already being developed at academic labs, national labs, and is being commercialized by private companies all around the world. But it is perhaps not too late to apply some restrictions, if necessary. Let us examine some of the dangers and ethical concerns that are brought up in relation to nanotechnology. The concerns listed in Table 23 are not my inventions, but they are representative of potential problems as expressed by thought leaders and nanotech critics.

We will see that many of the more far-reaching problems are not unique to nanotechnology, but come about because of the convergence of nanotech with many rapidly advancing fields, particularly biotechnology and information science.

Table 23 Societal concerns related to nanotechnology development.

1. Grey Goo Scenario – Environmental disaster due to self-replication
2. Green Goo Scenario – GMO organism takes over the world
3. Environmental disaster due to inhaleable or ingestible nanoparticles
4. Nanotech will end shortage-based economics
5. People will live for ever, leading to overpopulation
6. Only rich people will live forever; nanotech benefits will be unequally distributed
7. Nanotech will turn us all into cyborgs
8. Nanotech can be used to create incredible weapons of mass destruction
9. Nanotech will lead to machines that are smarter than we are
10. Nanotech will hasten the arrival of the Singularity

The Grey Goo Scenario

The "Grey Goo Scenario" is the appellation given to Drexler's fears of replicating nanobots (as discussed in Chapter 2). Drexler saw that, as technology developed, nanoscale robots could be developed that were essentially artificial life forms. He is not alone in this view. "Self-assembly and replication, the paradigms of molecular and cell biology, are being increasingly seen as desirable goals for engineering," said Phillip Ball (quoted in [3]), a journalist for *Nature*. "The alternative – laborious fabrication of individual structures 'by hand' – is still the way that electronic and micromechanical devices are made today, by a sequence of deposition, patterning, etching or mechanical manipulation that becomes ever harder as the scales shrink and the device areal density increases."

Self-assembly is a defining property – perhaps *the* defining property of life. A continual increase in the complexity of self-assembly, likewise is characteristic of evolution. Surely, the baroque method of reproduction used by human beings could not have been envisioned at the time the first primitive cells began replicating in the primordial ooze, three to four billion years ago.

As any molecular biologist will tell you, any bacteria, any eukaryotic cell, any organism – even a human being – is essentially a marvelously contrived machine. Even the ability to read these words is a consequence of tiny moving parts, protein molecules, electric currents, and nanofluidics.

An intelligently designed nanotech agent, however, with the properties of self-replication might escape into the general environment and wipe out the biosphere by sequestering to itself certain elements or materials necessary for life. It should be pointed out that any organism with a sufficiently short doubling time is, in theory, capable of taking over the planet, if left alone to do so. An *E. coli* bacterium with a replication rate of once every 20 minutes or so would create a lawn of bacteria covering the Earth within a few days, given sufficient resources. The reason that this does not happen is that there is effective competition for scarce resources among all life forms. Over evolutionary time, a kind of equilibrium of parity has been established. Every organism has a niche into which it fits, and from which it rarely crawls.

However, a nanotech agent, designed not by Nature but by a human being, perhaps with help from a super-computer and made from different materials than most life forms, might possibly have an extreme evolutionary edge over natural life forms. This is not a trivial threat.

As described earlier (see Chapter 2), Richard Smalley has argued that a nanorobot assembler of the kind originally envisioned by Drexler is not feasible due to problems with "fat fingers" that would be too large to manipulate matter in molecular spaces, and "sticky fingers" that would not be able to let go of an atom once it reached its destination. Drexler has intimated that Smalley was denying the case for assemblers in order to protect funding for the Nanotech Initiative. "Continued attempts to calm public fears by denying the feasibility of molecular manufacturing and nanoreplicators would inevitably fail," wrote Drexler, "thereby placing the entire field calling itself nanotechnology at risk of a destructive backlash. A better course would be to show that these developments are manageable and still distant [3]." James von Ehr pithily observed, "Eric didn't do himself any favors by getting into a pissing match with a Nobel-prize winner."

Actually, it is difficult to tell what kind of assembler it is that Smalley says cannot be built. The reflexive response of Drexler and his followers to any criticism of the concept of mechanosynthesis has been to direct the critic to Drexler's magnum opus, *Nanosystems: Molecular Machinery, Manufacturing and Computation* [4], as if the whole, total answer could be found in there somewhere. Aside from being a big, thick technical book that's a bear to read, nowhere in it does *Nanosystems* contain a blueprint for a molecular assembler. There are plenty of sketches of "molecular" gears, bearings, ratchets, manipulators and the like. Drexler's diagrams of molecular-scale objects look very much like engineering drawings of macroscale objects except where a scale is indicated. We are told, for instance, that a manipulator arm would involve 4 000 000 atoms, but we are not told which atoms, or how they would be put together. These boring details are apparently left for less imaginative engineers to fill in.

The debate over mechanosynthesis so far is huge to the participants, but mainly an entertaining academic diversion to most nanotechnologists. Because the description of a prototype mechano-synthesizer is so vague, it is a debate that is very difficult to handicap for the observer. Nevertheless, both sides would admit that the most primitive bacterium is a molecular assembler. If nanotechnology is pursued diligently by world researchers there should come a time when we are able to reverse-engineer a bacterium to the point where we are able to design a similar, though possibly inorganic, machine that can manufacture itself – the replicating nanobot. It may not turn out to be very much like the diamondoid nanorobot of Drexler's dreams, but it could still be a dangerous thing.

Two options present themselves dealing with the prospective threat of replicating nanobots:

1. Don't build them.
2. Keep them contained.

The first option holds the most safety. After all, though robots may be employed in the construction of other robots, there is no requirement that any single robot be a Universal Constructor. Robots, like cars, are built on an assembly line, with fixed robots doing individual jobs. This is actually more much more efficient than a robot replicating itself. The same would hold true at the nanoscale. The only current commercial plan for mechanosynthesis at the nanoscale is being developed at Zyvex, and will employ an assembly-line like process, although the project is still in early development.

The protocol for containment of replicating nanobots, if they are built, would likely mimic that already established for highly pathogenic viruses. Laboratories run by the Department of the Defense and Centers for Disease Control (CDC) have so-called "BSL-4" (biological safety laboratory 4) containment facilities to deal with certain pathogens, like Ebola virus. Containment involves fume hoods and laboratories that are under negative pressure so that no unfiltered air leaks into the environment. People never handle containers with virus in them directly, but must do so with rubber gloves that are built into hoods or with some sort of remote-control device.

It should be mentioned that accidents do happen. Long after smallpox had been eliminated from the world by an aggressive policy of vaccination, a technician in Great Britain was infected while working with one of the few remaining laboratory stocks. Several cases of SARS infection have been reported in Asian labs working with the virus. Containment in these labs was probably not as effective as it is in CDC labs, but that begs the point. Nanotechnology is a worldwide phenomenon; could we expect containment to be 100 % effective all the time?

Another aspect of containment for biological organisms has been to engineer them such that they would not be capable of survival outside the lab. For a bacterium, this might mean eliminating certain enzyme systems such that the organism would be dependent upon relatively rare amino acids, and thus be at a competitive disadvantage outside the laboratory environment. Presumably, something similar could be arranged for replicating nanobots, make them absolutely dependent on finding sufficient germanium arsenide to survive, for instance.

K. Eric Drexler warns darkly, “Runaway replication would only be the product of a deliberate and difficult engineering process, not an accident.” Grey goo, if it comes about, in Drexler’s view, would be through the misguided efforts of some future Dr. Strangelove or sociopathic terrorist.

Fortunately, replicating nanobots do not exist yet, and the technical skill required to design and build them probably does not exist yet either. So we have a little to time to consider the problem. It is easy to say: just don’t build them. But what authority could actually enforce that mandate in every laboratory all around the world? If the North Koreans, a small, economically disadvantaged country, can develop an atomic bomb, with all the complex infrastructure that weapon requires, how can we keep nanobots out of the hands of tin-pot dictators?

The Green Goo Scenario

By analogy with Grey Goo, we have Green Goo – the idea that a DNA-based artificial organism might escape from the lab and cause enormous environmental damage. This is more of a proximate danger, since of course, large numbers of genetically modified organisms already exist and some of them have been released into the wild. Since biotechnology is really “wet nanotechnology,” it is appropriate to consider the dangers of this practice in this chapter. Wet and dry nanotechnology are increasing converging into one field.

"Although the prospect of general assemblers may be quite distant, self-replicating ‘machines’ that use the tools of biology – and look more like living things than machines – might be closer at hand through the convergence of bio- and nanotechnologies,” notes Alexander H. Arnall, in a report prepared for Greenpeace [5].

Genetically modified organisms started in the 1970s. From an early stage, biologists recognized the danger and restrictions have been in place to keep organisms that were likely to be dangerous – modified pathogens, for instance – from escaping. Laboratory strains of *E. coli*, for example, are genetically crippled so that they will be at a competitive disadvantage to their wild-type counterpart. Laboratory disposal techniques are also designed to minimize the danger of escape, although it would be disingenuous to suggest that these are 100% effective. Human beings are the natural host for *E. coli* – we are essentially a perfect walking incubator for these bacteria.

The threat of Green Goo is somewhat mitigated because presumably any genetically modified organism would have to compete with its wild brothers, and most genetic modifications would put it at a disadvantage. This is not a guarantee, however. The problem is illustrated by the use of genetically modified crop plants. Certain grain plants have been modified with a "transgene" such that they produce a natural bacterial toxin that protects against insects. Such a modification might give a plant a competitive advantage. Crop plants are so heavily dependent on modern agriculture they do not usually survive well in the wild. But maize crops, for instance, have been shown to cross-breed with their wild cousins, transferring genetically modified transgenes in the process. Such a process of transferring competitive advantage to wild plants is unlikely to have catastrophic effects, but it may yield a particularly nasty, invasive weed.

During the past decade, a number of people have been interested in a "synthetic genome." The idea is to create an organism that has the minimal number of standard genes necessary for life. This could then be modified at will to create a patented made-to-order organism for specific purposes. Craig Venter, the genius behind the successful private effort to sequence the human genome, is interested in creating such an organism to lessen our dependence on hydrocarbon energy sources. Other uses of such an organism could be to manufacture biomedicines more efficiently than modified bacteria or yeast do, at present.

As discussed in Chapter 10, Venter has teamed up with Nobel Laureate Hamilton O. Smith to create such as an organism. As a proof-of-concept, they demonstrated that they could create a viable bacterial virus, called phi X, using off-the-shelf DNA precursors. Another team of scientists created a poliovirus from scratch. Since viruses are only metabolically active after infecting a host, this is not exactly like creating life, but it is getting close. In principle, a team of terrorists with sufficient technical skills could create a smallpox virus using the published genetic sequence.

Experimentally, genetically modified viruses that code for biomedicines have already been injected into people so that drugs might be more conveniently delivered where they are needed. Similar vectors are used to transfer genes to correct genetic defects in human clinical experiments. Even HIV, the virus that causes AIDS, has been altered to serve as a genetic therapy for fighting cancer.

If a genetically altered virus were to invade human germ cells it might be able to integrate its genes within the human genome, thus adding new genes to the human complement, in the same way that retroviral genes have already entered the genome of a number of mammalian species, including humans.

Is this an unlikely scenario? Generally speaking, extreme care is taken to make sure that these genes cannot be passed horizontally by infection. But so far there has been relatively little thought given to the possibility that therapeutic genes might be transmitted vertically through the genome. Though this is certainly an improbable event, the existence of retrovirus genes in the human genome is an existence proof that such vertical transmission is possible.

If you are dying of cancer or a genetic disease, the last thing on your mind, surely, is whether that virus might end up in your germ cells. Individually, it is kind of a moot point. But as a society, is it something we should worry about? This is an ethical debate that may impinge upon some areas of nanobiomedicine. We will discuss the intentional modification of the genome below under the Cyborg question.

Green Goo should probably be thought of as more of a gradual rather than a catastrophic phenomenon. Think of how much people have already modified the biosphere. Huge areas of the planet are given over to the growing of a few in-bred monoculture crops. Large forests have been converted to tree farms that produce only one kind of tree. By gradual selection and breeding, we have created large populations of animals, like cows and dogs, who are completely dependent upon us for their livelihood. Until the past few years, however, we have not had the capacity to intentionally rearrange genomes to create designer animals, plants and micro-organisms. Over the next 100 years or so, it would not be unlikely to find that much of the natural flora and fauna of the Earth have been replaced by genetically modified versions – creeping Green Goo. There are those who would object to this both on ethical and aesthetic grounds. But do the benefits to society outweigh these objections? This is not an easy question.

Margaret Atwood, a fine writer of literary fiction, has anticipated such a Green Goo-like phenomenon in her book *Oryx and Crake* [6]. Her future world is occupied by curious creatures like "pigoons" – pig/baboon hybrids created to supply organs for transplant, or "chickienobs" – strange birds with a dozen wings each, the better to supply the market for hot wings. These are not such fanciful creatures; several biotech firms have already modified pigs with human genes in the hope of obtaining transplantable organs. Farm-bred turkeys are already such fat stupid creatures that they are little more than machines for converting grain into meat, a far cry from the wild birds from which they are descended.

Environmental Catastrophe due to Inhaleable or Ingestible Nanoparticles

Once released into the environment, nanoparticles cannot be reclaimed. They will blow around in the atmosphere, dissolve in the oceans and rivers, enter the soil, and possibly enter the food-chain.

On the face of it, this is a particular problem with carbon nanotubes, which are harder than diamonds, stronger than steel, and not biodegradable. They penetrate through tissues with ease. Companies like Frontier Carbon Corp. and Carbon Nanotech Research Institute are already gearing up to produce multi-ton quantities of nanotubes. Current uses exist for nanotubes in making stronger, lighter carbon-fiber materials and in electronics. The fear is that nanotubes or other nanoparticles will contaminate the air or waterways, resulting in large-scale environmental damage. Comparisons are made with the miner's black lung disease, from coal-tar, lung cancer caused by cigarette smoking, or mesothelioma caused by asbestos fibers. These concerns cannot easily be reasoned away.

We should realize, however, that carbon nanoparticles have been around in the environment as a product of incomplete combustion since Prometheus brought fire to mankind. Industrial nanoparticles are already produced in large quantities in the form of carbon black, essentially colloidal carbon, also known as acetylene black, channel black, furnace black, lampblack, and thermal black. This is used in tires, inks, lacquers, carbon brushes, electrical conductors, and insulating materials. Typically, carbon black comes in 10- to 40-nm particles of elementary carbon with various chemicals adsorbed to its surface. Incomplete combustion also results in the production of fullerenes, carbon nanotubes, and probably a lot of other structures of which we are not yet aware.

A room-mate of mine once worked in a Bridgestone Tire factory in Nashville, TN. Every day he came home and took a shower before going to bed. Every morning, his sheets were dark with carbon black that he had sweated out during the night. Workers in tire plants adsorb this nanoparticle into their skin and they can sweat it onto their personal clothing even weeks after they leave their employment. Despite this intimate and long-term exposure, it appears to do them no harm. Carbon black is apparently innocuous, but that is unlikely to be the case with all nanoparticles, or even with all carbon nanoparticles.

One clear benefit of nanoparticle research will be improvement in the knowledge about what sorts of nanoparticles already exist in our environment – both those that are byproducts of industry and those that nature provides. Anything that burns carbon fuels, a furnace or an automobile, for instance, is likely to release carbon nanoparticles, as byproducts of imperfect combustion, into the air.

Nature also supplies airborne nanoparticles in the form of viruses, pollen, dust, and organic debris of decomposing plants and animals. The J. Craig Venter Institute has launched a pilot project, The Air Genome Project, focused on mid-town New York City, to better understand the diversity of microbes in urban air. The air samples will be analyzed at the Venter Institute's Joint Technology Center, one of the world's leading DNA sequencing facilities.

"No one really knows what is in the air," said Paula J. Olsiewski, program director, Alfred P. Sloan Foundation, which is funding the Venter project, to the tune of $2 million.

I was fortunate enough to be present at an informational meeting held in Washington between nanotech industry trade group leaders and representatives of the U.S. Environmental Protection Agency (EPA) and the Food and Drug Administration held in 2003. The meeting was held as an exchange of ideas about what challenges the nascent nanotechnology industry might present to regulators. The presentation of the EPA was illustrative of the problems that the EPA faces. For instance, certain chemicals it considers hazardous requiring special care in handling, transportation, storage and disposal. The EPA categorizes chemicals, naturally, on the basis of their chemical formula. But consider the problem of carbon nanotubes. To the EPA in 2003, a carbon nanotubes was just carbon, the same as graphite, the same as diamonds, the same as carbon black. There was no special care required in handling and, for that matter, as toxicology had not really been performed, nobody knew whether nanotubes were dangerous, or not.

The EPA representative did his best to reassure the nanotech leaders that his agency was not about to do anything to stifle the growing nanotech industry. Trade group leaders, particularly Bo Varga of NanoSIG, argued vociferously that the EPA should indeed step in with serious regulation. They are afraid of public backlash if serious problems develop. What with very premature worries over nanobots and the like, the industry wants the public to be comfortable with nanotechnology as commercialization ensues. No one wants a repeat of the turmoil that agribusiness went through over genetically modified crops.

Barbara Karn, who is in charge of directing nanotechnology research at EPA, has asked researchers to determine the potential for nanoparticles causing harm to the environment. According to Mark Wiesner, professor of Civil and Environmental Engineering at Rice, tests have shown that nanoparticles penetrate living cells and accumulate in the liver of experimental animals. He is especially worried about fullerene derivatives, like carbon nanotubes, which are extremely stable and therefore could be expected to accumulate in the environment over time.

Vicki L. Colvin, who is the executive director of the Center for Biological & Environmental Nanotechnology at Rice University (a nano-tech powerhouse) has repeatedly pointed out that there has been almost no research into the potential toxicology of nanoparticles. Colvin, however, is a definitely a supporter of the industry.

The U.S. government, to its credit, is using a good deal of the money allocated for nanotech to study possible environmental or health problems associated with nanoparticles. Purdue University scientists alone have recently been awarded over $2 million from the National Science Foundation and the EPA to study the environmental fate of carbon-based nanoparticles.

Unfortunately, we will still most likely deal with nanotech environmental hazards in the way that we always have, by ignoring them until they have become disastrous in an obvious way. Even that level of control will develop on a case-by-case basis. Robin Fretwell Wilson, of the University of South Carolina, School of Law, has pointed out that the EPA is set up by law as a reactive rather than proactive agency. The EPA exists to carry out environmental legislation passed by Congress. Such laws are passed usually only after serious problems are recognized in the environment, such as the DDT crisis, or the health problems in the wake of Love Canal pollution.

Nanotech Will End Shortage-Based Economics

Extreme nanotech enthusiasts expect that nanotechnology will usher in an era of plenty like nothing the world has ever known. Everything that you might desire, they say, from a Cadillac to a nicely grilled steak will be easily assembled from molecular feedstocks. All you will need is the right software and a molecular assembling device.

You might think that everyone would be willing to endorse such a future, but you would be wrong. I have heard it argued seriously at conferences that nanotech is a threat to society because it would end the shortage-based economics upon which the capitalist system is built. Imagine the social chaos that would result! There would be no point in being rich anymore. People would lose all motivation to do anything (this may be a peculiarly American point of view).

Let me state that I do not take either the extreme utopian or dystopian views of nanotech seriously. I am willing to argue, however, that what the world has now in the form of economic order is far from ideal. In many cases, "shortage-based economics" is a mirage. There is no shortage of food on the planet, yet a large percentage of the population goes hungry every night while industrial nations suffer from an epidemic of obesity. The problem isn't one of shortage; it's the method of distribution. Anything that nanotech can do to at least restore the necessities of life to those in need is, in my view, a wonderful thing. To say more would lead us off into political tangents that are better not discussed here.

People Will Live for Ever, Leading to Overpopulation

"The world becomes full of organisms that have what it takes to become ancestors."
River out of Eden, Richard Dawkins, 1995.

People die. It has always been thus. But not everyone is sanguine about the prospect of his own death. As Woody Allen so famously remarked, "I don't want to be immortal through my works; I want to be immortal through not dying."

Will nanotechnology, as Ray Kurzweil imagines, extend lifespans indefinitely? This is another one of those nanotech issues where one person's important feature is another person's bug in the program. Nanomedicine may progress to the point where we can fix all diseases of aging so that people will outlive their usefulness, but refuse to move on to their reward. Meanwhile, the younger generation will struggle and the world will become overpopulated. This is exactly the sort of world described in Bruce Sterling's book *Holy Fire* [7]. Long-lived "post-humans" become wealthy and dominate society, while under-employed younger folks dabble with anarchy.

Planned obsolescence is a design feature of the human body. Although we may reasonably expect to live the biblical "three-score years and ten," in practice, forty years is about as long as we can hope to remain in peak physical condition. Our best athletes give out by then. Sure, Nolan Ryan still threw 100 mile per hour fastballs into his forties and John Elway led the Denver Broncos to a Super Bowl victory at age thirty-seven. But these are the exceptions that prove the rule. Not many professionals, not even steroid-enhanced baseball players, can hope to maintain major league status so long.

Forty years, not coincidentally, is a generation times two. It takes about twenty years for humans to reach maturity and another twenty years to raise the next generation. Evolution has neatly arranged that we should last just long enough to train our own replacements.

Insuring our own posterity is a responsibility we tend to take very seriously. As usual, the renowned socio-anthropologist William Shakespeare said it first and best in his Sonnet II:

When forty winters shall besiege thy brow
And dig deep trenches in thy beauty's field,
Thy youth's proud livery, so gazed on now,
Will be a tatter'd weed, of small worth held:
Then being asked where all thy beauty lies,
Where all the treasure of thy lusty days,
To say, within thine own deep-sunken eyes,
Were an all-eating shame and thriftless praise.
How much more praise deserved thy beauty's use,
If thou couldst answer "This fair child of mine
Shall sum my count and make my old excuse",
Proving his beauty by succession thine.

Within the last decade, a number of women over the age of sixty have given birth. A torrent of controversy has developed over whether anybody should have children at such an advanced age, just because they can. Ignoring, for the moment, arguments on either side, one must be impressed with the incredible optimism demonstrated by the women involved. At an age when people used to be concerned with securing burial plots, these ladies are committing themselves to the care and feeding of another generation. They are presuming a high-functioning lifespan for themselves of eighty years, at least.

In-vitro fertilization, or "assisted reproduction," used to be a means of helping women and their partners achieve what most people took for granted, the opportunity to have a baby. Blocked fallopian tubes were probably the most common reason that most women were infertile. For the women involved and their partners, it was a personal tragedy, but it had no larger implications. However, until recently, *all* women were infertile after menopause. Until now, no one – not even the most privileged – had assumed reproductive rights after their ovaries had stopped functioning.

When the biblical Abraham and his wife Sarah were "well stricken with age and it ceased to be with Sarah after the manner of women," the Lord promised that Sarah would have a son. Sarah was so perplexed that she laughed at the thought of it. The Lord was a little miffed at Sarah's lack of faith. "Wherefore did Sarah laugh, saying Shall I of a surety bear a child, which am old?" He asked, "Is anything too hard for the Lord?" Is anything impossible with a sufficiently High Technology?

An indefinite lifespan, brought about by nanotechnology and other means, is certainly an event that Nature never intended. Aging is built into our genome through the pressure applied over time by evolution. By manipulating genetics, scientists have already been able to construct long-lived mice and fruit flies, and have more than tripled the lifespan of a certain nematode worm. That shows that aging is not an irresistible force, and that some plasticity exists with regard to aging, if only we knew how to control it.

What sort of effect would extended human lifespans have on population and social structure? Worries about overpopulation has been with us since Thomas Malthus [8] apparently invented the concept in 1798. What constitutes overpopulation turns out to be a culturally relative concept. The Indian people have successfully resisting forced sterilization and less coercive measures by the state to control population. This already crowded nation is on track to add 50% to its population by 2050 according to a report by the U.S. Census Bureau. On the other hand, Europeans have reduced their fertility to the point that some countries are well below replacement values. Italy, once renowned for its large happy families, now ranks among the lowest in the world with respect to fertility. Italians are practically an endangered species.

Serious people now worry about a crash in human populations. Such a crash appears to be well in progress in sub-Saharan Africa where life expectancies are back under forty years – an incredible reversal due to the AIDS epidemic. New epidemics, due to Nipah virus, the SARS virus, avian flu, Ebola or a pathogen yet to be discovered are threatened constantly in the headlines. This is not to suggest that overpopulation will not be a threat at some point in the future, but estimates of world population growth have moderated to the downside in recent years.

Population concerns aside, the more relevant question is whether nanobiomedicine can fulfill the expectation of adding substantial longevity. After all, modern medicine to this point has added significantly to the average life expectancy of those in industrialized countries, but precious little to the prospective longevity of any individual. The bell curve has shifted, but the outer bounds remain the same. People can live a productive life for their biblical three-score years and ten; maybe they can be active into their eighties, but still only a lucky few reach the century mark.

Nanomedicine, in my view, is likely to contribute to substantially greater longevity, in combination with other technologies, such as therapeutic cloning and tissue engineering that seem to have even a greater current potential for life extension.

Therapeutic cloning is the practice of generating patient-specific embryonic stem cells by transferring the nucleus of one of the patient's cells into an enucleated human egg cell. The egg cell is then allowed to undergo several divisions, forming an embryonic state called the blastocyst. Cells taken from "inner cell mass" of the blastocyst are called embryonic stem cells. They are "stem" cells because they have the potential to develop into any of the body's cell types or tissues. Although inhabited by an "old" genome, these cells have become new again, and have the same number of doublings available to them that an authentic embryonic cell would have.

Embryonic stem cells are being promoted as potential cures for diseases, like type I diabetes or Parkinson's disease, which result from the death of specific cell types. In principle, they could also be used to patch the injured portions of a person's heart after a myocardial infarction. Because they are patient-specific, they do not induce immunological responses. In essence, they would enable a whole new type of medical practice that relies on replacing cells rather than administering drugs, with all their side effects.

Therapeutic cloning has already engendered a fire-storm of political controversy. Ostensibly, this has been about the necessity of destroying "embryos" to create patient-specific embryonic stem cells. There is no fertilization involved between a sperm and egg cell involved in the process of therapeutic cloning, and therefore no creation of a new individual, just reinvigorated embryonic stem cells of the original individual. But the destruction of an embryo – no matter that it is only a ball of cells at this point – is seen by some, mostly the extreme American Christian right wing, as tantamount to murder.

Even if you do not subscribe to this extreme conservative view, it is true that the process has the potential to violate what has been termed "normative medicine." Normative medicine seeks to return the patient from a condition of disease to one of "normal" health. To add attributes or performance beyond what could be considered normal is not ethical medicine, according to one view.

Of course, medicine as practiced is already full of violations of normative medicine. Some are rather trivial, like Botox injections to remove signs of aging. Others, like sex-change surgery, are really drastic. The physician of the future will introduce changes in the patient that render him or her supernormal, at least in the judgment of the patient. We are well past the point of no return. Postmenopausal pregnancies have nothing to do with normality. Breast implants, face-lifts and liposuction are not about restoring health; actually, they are health-endangering for the purpose of lifting the patient to an idealized form. Sports medicine, both licit and illicit, seeks to equip our heroes with extraordinary abilities and endurance. We are not content with the human condition anymore – we yearn to be more. If it is possible, for instance, to replace heart tissues with new cardiomyocytes that behave as if they are newborn, then medicine is no longer normative. We are seeking to create something that has never existed before, an aging human with a newborn heart.

Nanomedicine has so far yielded new modes of drug delivery, imaging and diagnostics; benefits that have not been challenged on the basis of ethics. However, it is likely that as our control of matter at the nanoscale level increases, our ability to repair aging tissues will likely increase as well. The future may well include something like the mythical tiny submarines that voyage through our physiological systems, repairing cells as they go.

A nanotech-enabled artificial retina, as yet a primitive instrument, could be constructed such that people could see into the infrared or ultraviolet ranges. Such an individual might have certain advantages, as a soldier in battlefield situations, for instance. Thus, capabilities introduced by nanobiomedicine will eventually affect the debate over "normative" medicine as an ethical ideal.

Ray Kurzweil may be premature with his program to "live long enough to live forever." However, if you are not yet middle-aged, it seems likely that nanomedicine will contribute, along with other technologies, to the extension of your life beyond the historical limits for the human species, provided that you don't smoke, don't eat excessively, and don't take up sky-diving. Possibly you will be healthy and active long beyond the century mark; would you complain about that? Would you refuse treatment, on the grounds that such a lifespan is excessive, unethical and unfair to all the younger people? After all, Methuselah lived over 900 years, if the bible can be credited. Most people, I feel, would not object to treatment.

Only Rich People Will Live For Ever

Nanomedicine and other nanotechnology will surely benefit the rich first. That is the way of things in this world. A recent headline in the New York Times (May 16, 2005) read, *Life at the top in America isn't just better, it's longer.* "Class is a potent force in health and longevity in the United States. The more education and income people have, the less likely they are to have and die of heart disease, strokes, diabetes and many types of cancer. Upper-middle-class Americans live longer and in better health than middle-class Americans, who live longer and better than those at the bottom. And the gaps are widening..." wrote reporter Janny Scott. The article went on to describe disparate outcomes to a heart attack that varied by class.

The rich are the "first adopters" – the first to enjoy any new high-tech toy. Joe Six-Pack hungers after the new, flat, widescreen, high definition television, but cannot afford one. So he watches the Superbowl over at his brother-in-law's house, the son-of-bitch rich lawyer he can't stand. This is Ronald Reagan's "trickle-down" theory of economics in action.

Nanotech is part of a much larger debate about who controls and who benefits from new technology.

Some argue, for instance, that since nanotech research is being paid for by taxpayers, there should be some mechanism to spread the benefits in an equitable fashion. A gaping hole in this logic, to date, is that the more visible benefits of nanotech have actually been the result of private investment. The atomic force microscope is the result of efforts at IBM's Zurich laboratory. Carbon nanotubes were an accidental observation of a researcher at Nippon Electric Company. Dendrimers came from years of research at Dow Chemical.

New advances in medicine have always benefited the rich first. Implantable defibrillators, to cite a recent example, are incredibly expensive items, especially when surgical expenses are considered. Neither Medicare nor most insurance carriers will pay to implant them in all that could potentially benefit; nor will they pay for the most advanced models of these devices. However, Medicare will pay to implant these devices in certain patients, such as people with arrhythmia who have already suffered a heart attack. At least some patients who could not afford these devices from their own resources are granted the benefits of this technology. As technology improves, the bells and whistles in the advanced devices eventually become standard on all models. This "trickle-down" approach to medical distribution is not ideal, but it is the system currently in place, at least in the United States. Other countries with socialized medicine are arguably more equitable in the distribution of medical services. However, in countries such as Canada and Great Britain and even in formerly communist Russia, expensive private medicine thrives in competition with the much cheaper public service, which shows that some people are willing and able to pay up for premium care not available publicly.

A recent report from British academic societies on nanoscience and nanotechnologies [9], makes a further point:

"The application of science, technology and engineering has undoubtedly improved life expectancy and quality of life for many in the long term. In the short term, however, technological developments have not necessarily benefited all of humankind, and some have generated very definite 'winners' and 'losers'.

Concerns have been raised over the potential for nanotechnologies to intensify the gap between rich and poor countries because of their different capacities to develop and exploit nanotechnologies, leading to a so-called 'nanodivide'. If global economic progress in producing high-value products and services depends upon exploiting scientific knowledge, the high entry price for new procedures and skills (for example, in the medical domain) is very likely to exacerbate existing divisions between rich and poor."

Table 24, which is taken from an article titled *Nanotechnology and Life Extension* by Patricia Connally [10], illustrates one reason that nanotechnology is likely to deepen the divisions between developed and underdeveloped nations with regard to medical care. Stated plainly, the medical problems of the two groups are vastly different. Although the problems of underdeveloped nations are arguably much more severe (consider the loss of life caused by infectious diseases, for instance), nanotechnology will be directed, at least initially, to solve the problems of the society in which the technology is being developed.

"The problems of the developing world are quite different, and it might be argued that unless life extension in this environment is addressed by those who have the technology and wealth to do so, then the stability of developed societies worldwide will be affected," says Connally, although she does not specify the mechanism of instability. The implication is that the poor would rise up against the rich; recent history suggests the more likely scenario is a flood of illegal immigration from technology-poor nations to technology-rich nations.

Ideally, technology should benefit all who can benefit; however, the lack of a societal mechanism to insure such a distribution should hardly be regarded as reason to delay the development of the technology, in the first instance. Neither nanotechnologists nor the institutions that employ them bear any special responsibility for the organization of society.

Table 24 The challenges to life extension in developed and developing countries.

Target groups	Quality of life problems	Major causes of death and disability
Developed countries: Aging populations only	Loss of strength and mobility Loss of mental sharpness/ neurological disease Social isolation Poverty	Cardiovascular disease Diabetes and its complications Inflammatory diseases, including arthritis Cancer Neurological disease or impairment
Developing countries: All age groups	Environmental, lack of safe water and sanitation Disease related loss of earnings Poverty Malnutrition	Infectious diseases Parasites Cardiovascular disease

Source: Patricia Connally [10].

Nanotechnology Will Turn Us Into Cyborgs

This objection can be restated to read, "Nanotechnology will accelerate the trends that are turning us all into cyborgs." Resistance has really been futile for some time.

The word "cyborg" or cybernetic organism, was originally coined by Manfred Clynes and Nathan Kline in 1957 to refer to humans equipped with mechanical devices and/or physiological alterations that would allow them to survive the rigors of space travel [11]. NASA carried the idea forward in 1963 with its Cyborg Study to consider the "theoretical possibility of incorporating artificial organs, drugs, and/or hypothermia (low temperature) as integral parts of life-support systems in space craft design of the future." Though few cyborgs have yet to escape the Earth's atmosphere, they have proliferated on the ground at a furious rate.

Cyborgization, though not recognized as such, began slowly and innocently enough with medical necessity, long before NASA existed. The body has been invaded by bits of hardware, inconsequential individually, but perhaps alarming in their totality. For hundreds of years, spectacles have corrected poor vision; now contact lenses do so more intimately. George Washington complained about his wooden teeth. Many of his soldiers came back from the Revolutionary War with wooden legs. Plastic hip bones and shoulder sockets are now commonplace among the elderly. Silicone breasts replace the sometimes disappointing endowments of Nature. Dialysis machines functionally replace defective kidneys. A Los Angeles surgeon, Achilles Demetriou has invented a cyborg liver; a combination of filters and *ex-vivo* hepatic cells. The chest cavity was penetrated first by the pacemaker and more recently by the artificial heart. The brain is no longer a virgin;

cochlear devices for the hearing impaired stimulate the auditory nerve electronically. Nanotech electronic retinal implants are in the works as a visual prosthesis for the blind.

On an extra-organismal level, many – if not most of us – have become tethered to a computer or at least a telephone to accomplish our daily tasks. Some of us have been known to circle a parking lot for 30 minutes to avoid walking an extra hundred yards, substituting the automobile for legs. Because of paraplegia, breathing difficulties, or extreme weight, many people require wheelchairs for locomotion.

The figure of the cyborg has been expropriated (some might say misappropriated) from male-dominated sci-fi action flicks by culture theorist Donna Haraway [12] for use as a metaphor and symbol of her own brand of feminism. Her shape-shifting cyborg is in revolt against its creator, the allegedly capitalist, racist, sexist, military/industrial society. In effect, she turned the Terminator against the few male chauvinists who survived the political onslaught of the previous decades. These wary few might have been forgiven if they heard a radical feminist cackling in the wings when well-intentioned veterinary researcher Ian Wilmut cloned a female sheep. This animal, the genetic material of which was derived entirely from the nucleus of an udder cell, was named Dolly in honor of the country star, Dolly Parton, and her mammaries. This extravagant use of technology has made assisted parthenogenesis possible; the male of the species is now completely dispensable.

But ladies, the war of the sexes is far from over. Two male biomedical researchers, Robert Langer of MIT and Joseph Vacanti of Harvard Medical School, have publicly predicted the imminent development of an artificial womb. Finally, reproduction will be completely dissociated, not only from sex, but from motherhood. From the day the first Cesarean section was performed, this moment was foreordained. Cyborg babies are on the way. The crying newborn cyborg will be comforted by the soothing hum and gurgle of its former habitat, the artificial womb, just as today's newborn baby is quieted by the nearness of his mother's heartbeat.

Body parts are now being engineered. The world was astounded some years ago by a picture of an apparent human ear grown on the back of a laboratory mouse. The project, though largely the work of a MIT chemical engineering professor, Linda Griffith Cima, was suggested by a colleague, a plastic surgeon who wanted to have such a product for use in reconstructive surgery. Vital organs are also being redesigned. If you're not happy with the idea of replacing your aging heart with an electric pump and you can't find a willing human donor, a number of biotech firms are developing an alternative. How about a pig heart? Pigs have been seeded with human genes by American biotech firms such as Nextran or Alexion Pharmaceuticals in the hope of creating universal organ donors. So far, it is still unclear how many human genes will be required before a pig heart is acceptable to a human body. Another question to consider: How many human genes must a pig have before he can sue for civil rights violations?

Or, instead of adding mechanical parts to a human, how about adding human parts to a machine? A group of scientists at UCLA cultured muscle cells between gold points held in a silicon frame. When the muscle fibers were completed, they released their hybrid creature and it crawled away. They imagine the day when such muscle-powered robots could function within our bodies, living off the glucose in our blood, a mechanical parasite, hopefully one that they could endow with a symbiotic function, such as shoveling the plaque from our arteries.

The U.S. National Science Foundation, in collaboration with the Defense Department, has issued a 400-page plus report called *Converging Technologies for Improving Human Performance: Nanotechnology, Biotechnology, Information Technology and Cognitive Science* [13]. The NSF argues persuasively that these technologies, which it abbreviates to NBIC, are converging based on the "material unity at the nanoscale and technology integration from that scale" Basically, since everything is composed of atoms and molecules, all technology that works at that scale dissolves into one discipline, at least insofar as design, manufacturing and engineering processes are concerned. "At this unique moment in the history of technical achievement", says the NSF, "*improvement of human performance through integration of technologies* becomes possible." What do they mean by that? Their list of payoffs includes "enhancing individual sensory and cognitive capabilities, revolutionary changes in healthcare, improving both individual and group creativity, highly effective communication techniques including brain to brain interaction, perfecting human-machine interfaces including neuromorphic engineering, enhancing human capabilities for defense purposes ... ." In other words, they want to make telepathic, superhuman cyborg soldiers – like Jean-Claude van Damme in the *Universal Soldier.* And you wonder why the Bush Administration bought into the National Nanotech Initiative? At $3.7 billion, it's a bargain.

The environmental group ETC has satirized the NSF's converging NBIC as BANG, for bits (information technology), atoms (nanotechnology), nerves (cognitive science) and genes (biotechnology), designations apparently filched from a James Canton figure contained in the NSF report [13]. "BANG will allow human security and health – even cultural and genetic diversity – to be firmly in the hands of a convergent technocracy," warns ETC.

Do you think that this radical group is unduly paranoid? Consider some of the NSF/DOD bullet points:

- Fast, broadband-width interfaces directly between the human brain and machines will transform work in factories, control of automobiles, ensure superiority of military vehicles, and enable news ports, art forms and modes of interaction between people.
- Comfortable, wearable sensors and computers will enhance every person's awareness of his or her health condition, environment, concerning potential hazards, local businesses, natural resources and chemical pollutants.
- Robots and software agents will be far more useful for human beings, because they will operate on principles of human goals, awareness, and personality.
- National security will be greatly strengthened by light-weight, information-rich war fighter systems, capable uninhabited com-

bat vehicles, adaptable smart materials, invulnerable data networks, superior intelligence gathering systems, and effective measures against biological ,chemical, radiological, and nuclear attacks.
- The ability to control the genetics of *humans* (italics added), animals, and agricultural plants will greatly benefit human welfare, in accordance with a widespread consensus about ethical, legal, and moral issues.

Well, one person's consensus is another person's controversy. I have argued publicly that the cyborgian fantasies emanating from the NSF in *Converging Technologies* were essentially a lobbying effort designed to appeal to the military-oriented Bush Administration, to make sure that Bush, followed through on the National Nanotech Initiative begun in the Clinton Administration. If so, it was successful; the $3.7 billion 21st Century Nanotechnology Research and Development Act was passed and signed. It must be admitted, however, that many outstanding scientists contributed to the NSF/DOD report, as well as a few unlikely luminaries, such as right-wing politico Newt Gingrich and pop sociologist Sherri Turkle. And editor Mike Roco, when questioned, specifically rejected the idea that the Nanotech Initiative was purposefully skewed toward military objectives.

One of the major visionary projects is to develop a direct brain–machine interface, so that people can communicate with computers without the slow and annoying interface of keyboard or speech.. This would also allow direct brain implants. In William Gibson's classic science fiction novel *Neuromancer* [14], these implants allowed, for instance, the instant mastery of new subjects, like a foreign language, nuclear physics or safe-cracking. Gibson called his implants *microsofts*; in NSF's updated vision these implants might be called *nanosofts*. The reduction in scale vastly increases the number of individual neurons that can be addressed.

"We propose to develop a new technology that would allow *direct* interaction of a machine with the human brain and that would be secure and minimally invasive", say Rudolfo Llinas and Valeri Makarov, of New York University Medical School [13]. Their idea is to use the vascular system as pipeline to neural connections. "The capillary bed," they point out, consists of 25 000 meters of arteriovenous capillary connections with a gage of approximately 10 microns." Plenty of room in those capillaries for nanoconnections.

Miguel A. L. Nicolelis of Duke University Medical Center and Mandayam A. Srinivasan, MIT, talk about the advantages of a brain–machine interface. One of these is "scaling of position and motion," so that a "slave" actuator, being controlled directly by the subject's voluntary brain activity, can operate within workspaces that are either far smaller (e.g., nanoscale) or far bigger (e.g., space robots; industrial robots, cranes, etc.) than our normal reach.

Nicolelis is the principal investigator of a $26 million contract to Duke University sponsored by the Defense Advanced Research Projects Agency (DARPA). The contract is part of DARPA's Brain-Machine Interfaces Program that seeks, in part, to develop new technologies for augmenting human performance by accessing the brain in real time using the information to drive external devices.

In 2000, Nicolelis and his colleagues tested a system on monkeys that enabled the animals to use their brain signals, as detected by implanted electrodes, to control a robot arm to reach for a piece of food. The scientists even transmitted the monkey's brain signals over the Internet, remotely controlling a robot arm 600 miles away.

So, does nanotechnology promise to accelerate the trend toward cyborgization? You bet, and American tax dollars are already hard at work toward that end.

To quote the *Converging Technologies* report once again:

> "It is hard to find the right metaphor to see a century into the future, but it may be that humanity would become like a single, transcendent nervous system, an interconnected 'brain' based in new core pathways of society."

Uncle Sam has you in his sights. Resistance is futile. You will be assimilated.

Nanotechnology Could Create Weapons of Mass Destruction

Nanotechnology has just begun as a separate discipline, and already the military has gotten into the act. For instance, MIT has established the Institute for Soldier Nanotechnology (ISN). The ISN focuses on six key soldier capabilities: threat detection; threat neutralization (such as combat armor, chemical weapons suits); concealment; enhanced human performance; real-time automated medical treatment; and reduced logistical footprint. Part of the latter involves just lightening the load of stuff that the infantry carries into battle. The U.S. army is hoping that building at the nanoscale will allow a host of functions and protections to be built into a standard combat outfit (Table 25).

Table 25 Nanotechnology for the "super warrior."

Protective clothing	Nanofibers, permi-selective membranes, nanoreactor coatings, chameleon-like color adaptation
Armor	Ballistic face shield (polymer-layered silicates); ballistic helmet (nanotubes, nanofibers, nanoparticulates)
Sensors	Chemical, biological, explosives detection
Power sources	Compact fuel cells
Clean water	Nanofilters
Enhanced weaponry	Light-weight composites

If militaries around the world were only worried about outfitting their infantrymen, it would not be so worrisome. But clearly they are interested in new and more powerful weapons.

DARPA's Augmented Cognition Program is a cyborgian affair that promises a "symbiotic marriage" between man and machine to extend the capabilities of war fighters. Among the things that DARPA would like to do is to extend the soldier's cognitive performance so that they can function for days without sleep. One modest proposal is to "stimulate the normal neurogenesis process that is part of learning and memory, thereby increasing the reserve capacity of the memory circuits"; in other words, grow more nerve cells. If that isn't enough, DARPA would like "brain machine interfaces to explore augmenting human performance by extracting neural codes for integrating and controlling of peripheral devices and systems." So, like your computer, the soldier will have "peripheral devices" that he can operate just by thinking about it. The M-16, the tank, fighter plane, the smart bombs will become extensions of the military man's personality.

The war in Iraq has been a proving ground for new robotic technologies. The U.S. has deployed an armed mobile robot that can be controlled from a distance. A camera allows a remote operator to aim and fire at enemies. Add a wireless internet connection and the robotic soldier could, in theory, be controlled by an operator sitting comfortably at his desk in the Pentagon. There has already been a commercial spin-off: a ranch in Texas offers hunting via the Internet. To outfox game regulations, they stock the ranch with exotic, non-native species not covered by local statutes. You track your prey via remote cameras. One click of the mouse kills it for you. For an extra fee, the ranch will stuff and mount the head and ship it to you Fed-Ex.

Isaac Aasimov's First Law of Robotics in his classic sci-fi novel I. Robot was that "a robot should not harm a human being or, through inaction, cause a human being to come to harm." In the real world of military robots, the first law has already been discarded.

"Intelligent weapons systems are already beginning to emerge. Cruise missiles, smart bombs, and unmanned reconnaissance aircraft have been deployed and used in combat with positive effect," notes James Albus [13]. "Unmanned ground vehicles and computer-augmented command and control systems are currently being developed and will soon be deployed. Unmanned undersea vehicles are patrolling the oceans collecting data and gathering intelligence. These are but the vanguard of a whole new generation of military systems that will become possible as soon as intelligent systems engineering becomes a mature discipline."

In the not too distant future, in other words, war will have become so automated that it won't require many soldiers.

"It is envisioned that in 20–30 years from now, when the research and development are successfully completed, nano-bio-info-cogno (NBIC) technology will enable us to replace the fighter pilot, either autonomously or with the pilot-in-the-loop, in many dangerous warfighting missions. The uninhabited air vehicle will have an artificial 'brain' that can emulate a skillful fighter pilot in the performance of its missions," says Clifford Lau [13], from the office of the undersecretary of the Department of Defense.

An armed, inanimate, intelligently controlled, flying enemy robot seems pretty intimidating. Now imagine that the robot was too small to see. Such invisible robots do not exist yet, but they can't be entirely ruled out. "Nanotechnology is expected to improve the performance of DOD IT systems by several orders of magnitude," notes Lau.

The fusion of biotechnology with nanotechnology could possibly create horrific new biologic weapons. Smallpox is only a good weapon if your population has been vaccinated and the enemy hasn't. Otherwise, the epidemic inevitably goes global and everybody loses. But suppose you could invent something like an Ebola virus with a time-clock; one that would self-destruct after a certain number of generations. Such a virus could be used with impunity, knowing that the epidemic would burn itself out before it got beyond enemy borders.

Another disastrous possibility would be use of nanotechnology in the creation of the so-called "fourth generation of nuclear weapons." The defining technical characteristic of these weapons is the triggering – by some advanced technology, possibly a superlaser – of a relatively small thermonuclear explosion in which a deuterium-tritium mixture is exploded in a device that is only a few kilograms in weight, small enough to put in a briefcase.

While nanotechnology is conquering the three-dimensional space, the fourth dimension of time has been under attack for some time. "... it should not be forgotten that while nanotechnology mostly emphasizes the *spatial* extension of matter at the scale of the nanometer (the size of a few atoms), the *time* dimension of mechanical engineering has recently reached its ultimate limit at the scale of the *femtosecond* (the time taken by an electron to circle an atom, a millionth of a nanosecond, for those of you keeping score)", says Andre Gsponer [15]. "It has thus become possible to generate bursts of energy in suitably packaged pulses in space and time that have critical applications in nanotechnology, and to focus pulses of particle or laser beams with extremely short durations on a few micrometers down to a few nanometer-sized targets. The invention of the 'superlaser,' which enabled such a feat and provided a factor of one million increase in the instantaneous power of tabletop lasers, is possibly the most significant recent advance in military technology." Gsponer has been echoed by Jane's Information Group, an international security consultant.

The military pursuit of nanotechnology, scary as it is, is not all bad. Much of the technological progress of the twentieth century was the indirect result of the search for military dominance. Atomic power, satellite communications, global positioning systems, jet aircraft and space travel all had military origins. The Internet is the direct descendant of ARPANET, a distributed communications system designed to withstand a nuclear attack.

"The predecessor to the Internet, ARPANET, wouldn't have occurred without two things," notes Newt Gingrich [13]. One was ARPA itself (now called DARPA), which had the funding, and the second was a vision that we should not be decapitated by a nuclear strike. People tend to forget that the capacity to surf on the Web in order to buy things is a direct function of our fear of nuclear war."

Nuclear war was a fear that my generation was brought up on, in the wake of the Second World War. "Duck and cover" drills were held in the schools. A siren would come over the school public address system and we were supposed to drop what we were doing and cower under our desks until we heard the all-clear signal. Every time I did this, in my mind's eye I saw the destruction of Nagasaki and a mushroom cloud. What protection could there possibly be, really, in the frail student's desk top above my head?

Entrepreneurs sold back-yard bomb shelters; many were later converted to potting sheds or darkrooms. For use in the event of war, my grandmother had a box of canned goods stored in her basement, the labels faded and peeling off. If war had indeed come, every dinner would have been a pot-luck experience.

But war – at least, the nuclear kind – did not come again in the twentieth century. We were spared by dumb luck, the grace of God or the mutually assured destruction (MAD) policies of the U.S. and the Soviet Union. Take your pick. Though I wouldn't discount the other choices, I am sure that MAD played a part. Nuclear war was an unwinnable proposition.

The military use of nanotechnology brings with it new terrors to a new generation. Surely, the military of every nation sees an obligation to keep from being on the losing end of new technology, as indeed they should. Part of the trick of surviving the twentieth century lay in maintaining civilian control over the military, at least in the advanced industrial nations. Technologically advanced weapons will not hurt us, so long as they are not used. Perhaps this is a feeble answer to fears about military nanotechnology, a metaphorical desktop to hide under. But it's the best I have.

Nanotech Will Create Machines that are Smarter than Human Beings

Could a mobile robot equipped with a quantum computer capable of learning be made more intelligent than a human being? Right now, this is an unanswerable question. Part of the problem lies in the definition of intelligence. Certainly we already have machines that are far more capable than we in performing calculations of almost any kind. Still, that intelligence is "brittle" to use the technical term. On the web, for instance, when a webmaster wants to keep a form from being filled out automatically by some sort of autonomous software agent, he includes letters or number codes that are distorted. A human being can usually discern the meaning behind these fractured fonts, but machines so far cannot. Some types of pattern recognition have been difficult to program. But it is doubtful that these barriers will stand for long.

The head of British Telecom's Futurology unit, Ian Pearson, in an interview with the UK Observer, pointed out that Sony PlayStation 3 is 35 times more powerful than its predecessor. A decade earlier it would have been considered a supercomputer. According to Pearson, the Playstation 3 is 1% as powerful as the human brain. Playstation 5 may be well beyond us. Pearson expects a "fully-conscious" supercomputer with above human intelligence to be built by 2020.

Can a computer be made conscious and self-aware? Gary Kasparov, who in 1997 was the best human chess player (some say the best human chess player ever), lost a chess match to the IBM machine Deep Blue. But Deep Blue did not then go out and celebrate the event. It cannot be said, really, that the machine even knew that it played a match. It just did what it did.

For an intelligent machine to become problematical, it must have autonomy: Like the computer HAL in the Stanley Kubrick movie *2001: Space Odyssey*, it must do things on its own authority.

The fear seems to be that if enough intelligence is built into a machine, it will become not only autonomous but, in effect, a human replacement. Can a machine really think? In 1950, the brilliant mathematician Alan Turing answered that question in the affirmative. How would we know that a machine is conscious? How do you know that other *people* are conscious? This is the gist of a Turing test: if you can hold a conversation with a computer without realizing that it's a computer, then for all practical purposes, it is conscious.

In 1990, Hugh Loebner agreed to underwrite a contest designed to implement the Turing Test to be conducted by the Cambridge Center for Behavioral Studies. Loebner pledged a Grand Prize of $100 000 and a Gold Medal or the first computer whose responses were indistinguishable from those of a human. So far, there have been no claimants, but each year an annual prize of $2000 (upped to $3000 in 2005) and a bronze medal is awarded to the computer program that gets the closest. A whole class of programs, called chatbots, has been built with the aim of imitating human conversation. The fifteenth annual event will take place in September, 2005.

The fear of machines that would replace men goes back to the Luddites (followers of the fictional Ned Ludd), who broke up weaving machines in the eighteenth century in protest of the jobs that were lost to automation. Fear of the thinking machine has a more recent origin, one so far rooted only in science fiction. It is too early to tell whether nanotech and quantum computing could make a conscious, autonomous thinking machine a reality. But given that nanotech could enable such a thing, it is doubtful that foregoing nanotech would keep such a thinking machine from becoming a reality. Nanotechnology would only add technical sophistication to HAL and his cousins.

We should keep it in mind that greater-than-human intelligence is not required to make a machine either autonomous or perilous. A hippopotamus is probably no smarter than a barnyard pig, but it is the most dangerous animal in Africa outside of humans. How do you tell if a hippopotamus is conscious? What difference does it make? If you see three tons of animal charging toward you, get out of the way!

The real threat from machines, if it comes, may be from those, like the hippos, which do not care to imitate humans. They just do what they do and if humans get in the way, this is not their concern.

Nanotech Will Hasten the Arrival of the Singularity

János Lajos Margittai Neumann was sort of an earlier version of Richard Feynman – a polymath of incredible vision. His father's purchase of German title of nobility caused a name change to János von Neumann (sometimes spelled van Neumann), later anglicized to John Von Neumann when he moved to the U.S. in the 1930s. He was an early proponent of game theory, and it was he who authored the "mutually assured destruction (MAD)" policy of the U.S. during the Cold War. As a mathematician, he provided a rigorous formulation of quantum mechanics. As recounted earlier, Von Neumann was the originator of the idea of the universal assembler, which K. Eric Drexler shrank to nanoscale, at least conceptually. He created the field of cellular automata, which through modern adherents like Stephen Wolfram, has become all-encompassing method for describing natural processes. And just as an aside, a kind of throw-away line, von Neumann predicted the technological Singularity. Actually, there is not even a written record of it, only hearsay evidence from von Neumann's acquaintance, Polish mathematician Stanislaw Ulam:

> "One conversation centered on the ever accelerating progress of technology and changes in the mode of human life, which gives the appearance of approaching some essential singularity in the history of the race beyond which human affairs, as we know them, could not continue."

Technology, as Ray Kurzweil has noted, follows the law of accelerating returns. There is a positive feedback loop by which advancing technology begets yet more technology and an ever-increasing rate of speed. To get an appreciation for this effect, it is only necessary to view Table 26, which recounts the cultural evolution of our species. Modern human beings have been extant for 200 000 years. Our brains and physiology have not changed much during that period, if at all. But almost all of what we would call science and technology is crammed into the last 100 years or so.

Table 26 Human cultural evolution.

Years ago	Key advancements
200 000	Tools
10 0000	Art
8000	Agriculture, writing, libraries
800	Universities
500	Printing
300	Renaissance in science and technology, accurate clocks
200	Industrial revolution

Years ago	Key advancements
100	Telephone, electric lights, automobile, air flight
80	Radio
60	Television, antibiotics, vaccines, atomic bomb, electron microscope
50	Computers, organ transplants
40	Space travel, molecular biology
30	Personal computer
20	Internet, cell phones, gene modified crops, cochlear implants
<20	Biotechnology products, cloned animals and plants, GPS for navigation, nanoscience, mobile robots, Internet, atomic force microscope

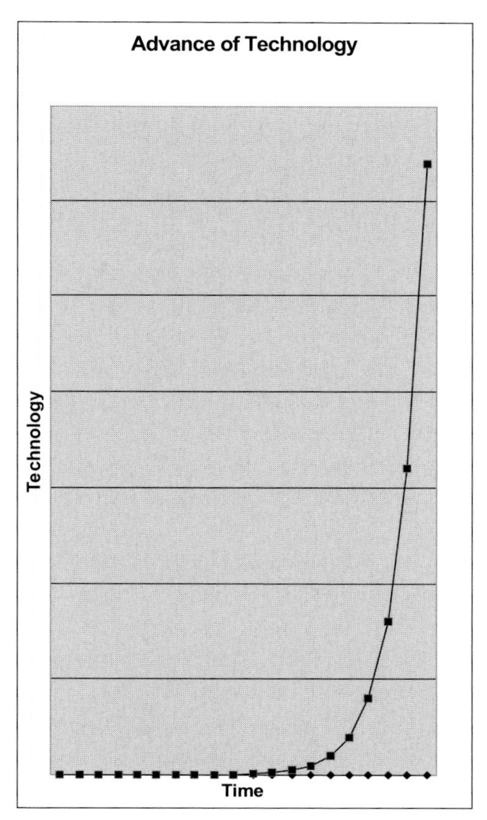

Figure 46 Exponential growth of technology leading to a Singularity? A plot of an exponential curve, 2^n.

If you could make a mathematical expression for the increase in technology, it would seem to take an exponential form. Following the curve leads ultimately to an asymptotic line pointing straight up (Fig. 46). This is the Singularity. Where it leads, say its adherents, is beyond the capability of humans to predict or understand.

Credit for popularizing the notion of a singularity goes to mathematician and computer scientist Vernor Vinge, who is also the author of such science-fiction classics as *True Names* and *Marooned in Real Time*. Just over the horizon, says Vinge, are changes so momentous that they are "... comparable to the rise of human life on earth. The cause of this change he envisions is "... the imminent creation by technology of entities with greater than human intelligence", an event he expects to occur between the years 2005 and 2030. With super-human intelligence in charge of technological progress, progress itself becomes much more rapid and involves the creation of still more intelligent "beings" on an ever-shortening time scale.

"We humans," says Vinge, "have the ability to internalize the world and conduct 'what ifs' in our heads; we can solve many problems thousands of times faster than (Darwinian) natural selection. Now by creating the means to execute those simulations at much higher speeds, we are entering a regime as radically different from our human past as we humans are from the lower animals.... From the human point of view this change will be a throwing away of all previous rules, perhaps in the blink of an eye, an exponential runaway beyond any hope of control." This eye blink – "the Singularity" is a time when change occurs at such blinding speed that mere humans will be rendered obsolescent – "the physical extinction of the human race is one possibility."

When it comes to information technology, things do have a way of getting out of hand in a hurry. Wozniak and Jobs started mucking around in their garage putting together mismatched components, and Presto/Chango, the personal computer industry was born. Relational database programs supposedly grew out of a desire to handicap football games. The World Wide Web was invented by Tim Berners Lee almost by accident as a way to share information with his physicist buddies. In 1993, there were all of fifty known HTTP servers. Suddenly thereafter, the Web just kind of happened, an emergent phenomenon that hardly anyone (other than Al Gore, of course) expected.

No doubt some geek, with the aim of automating his buying and selling on E-bay, will invent an artificial intelligence program that's just a little too smart. The thing will start buying and selling on Wall Street for its own account, and in so doing will finance a machine-dominated Singularity that renders us all obsolete before we know it.

Vinge is decidedly ambivalent about the Singularity: he thinks that it is unavoidable and therefore should be embraced. But he doesn't necessarily view it as a good thing. There are among us people who call themselves transhumanists – humans in transition to an uncharacterized post-human state. Some of these folks organize themselves into cult-like groups, for instance the Extropians, based in California. They view the coming Singularity as a kind of techno-rapture.

"We are about to enter a period of gargantuan change happening in an astonishingly short amount of time and that is an inherently dangerous situation, it would be foolish to deny it," said one avowed Transhumanist. "In spite of the dangers I admit I'm happy about the coming changes, we might survive it, and the alternative after all, is old age and death for all of us [16]."

"The technological expansion of my own intelligence may serve to keep me ahead of the game," said another, "such that I become the creator of the Singularity rather than it's victim [16]."

For Vernor Vinge, the Singularity is all about artificial intelligence, about computers who are "awake" (i.e., conscious) and super-humanly intelligent. "Machines are becoming more human," said the great science fiction writer Phillip K. Dick in a speech during the relatively innocent year of 1972. "Our environment, and I mean our man-made world of machines, is becoming alive in ways specifically and fundamentally analogous to ourselves." How far we have come since then. I wonder what Dick would say now, if he were alive?

Nanotechnology could certainly increase the possibility of "intelligent machines." Already, semiconductor chips have broken the 100-nm level for chip features. By this standard, nanotechnology is already increasing the speed and power of computers Nanotech may also result in the development of quantum computers that are orders of magnitude more powerful than the digital computers of today. If power alone is enough to "wake-up" a computer, then quantum computers will be more conscious than we are.

Ray Kurzweil is a newer, more enthusiastic proponent of the Singularity, and he includes a whole list of technologies that will contribute to the Event, including nanotechnology and biotechnology. Kurzweil's book, *The Singularity is Near: When Humans Transcend Biology*, is scheduled to be distributed in September, 2005. Kurzweil has described the Singularity as "technological change so rapid and profound it could create a rupture in the very fabric of human history [17]." He bases his projection on his Law of Accelerating Returns. According to Kurzweil, "An analysis of the history of technology shows that technological change is exponential, contrary to the common-sense 'intuitive linear' view. So we won't experience 100 years of progress in the 21st century – it will be more like 20 000 years of progress (at today's rate)."

There are, perhaps, exceptions to be found to the overall march of technology. One problematic area, for instance, has been the field of drug development. The 1990s should have been the golden era for drug discovery. New methods of "combinatorial chemistry" had made it possible to create almost any organic compound that could be imagined. The sequencing of the human genome was revealing new drug targets on a regular basis. Robotic technologies were automating the process of preclinical testing to the tune of 100 000 individual assays per day. Everything was in place for incredibly rapid progress. But somehow it didn't happen. At a drug discovery conference, held in June 2000, I heard the keynote speaker Doug Livingstone of Novartis Research Foundation moan, "We bought you the robots, now where are the drugs?"

The buzzword of the conference was "de-bottlenecking," an egregious neologism that implies a negation of the verb "to bottleneck (as in: I bottleneck, he bottlenecks, they are bottlenecking, we should be all be de-bottlenecking). When I was a kid, at parties I would spin a bottle between members of the opposite sex, who were circled around it. Whomever de bottle pointed to, her I would get to neck with. With all this de bottle necking going around, it is not surprising that little was being done in the realm of actual drug discovery.

After the robots were deployed, the slow point in drug development, by common consensus, was assay development. According to a number of companies, this problem could be fixed if you buy their: _____ (fill in the blank) (a) fluorescent protein; (b) electrochemiluminescent reagent; (c) ion channel technology; (d) microfluidic chips; (e) cell line; (f) software. All of which might have seemed true to a participant of the conference in June 2000. But the real bottleneck, it seems in retrospect, was a shortage of expertise. It takes a clever person (or two or three or four) to devise an assay that will give a meaningful result.

Another problem for the drug industry was the implementation of an innovative management concept called "drug champions." Researchers, you may be surprised to learn, reside toward the bottom of the totem pole when it comes to drug companies. They are hired and fired on a project basis while management is, relatively speaking, forever. The idea was that researchers who had discovered a new drug candidate would work their way up the management ladder by following their drug and serving as an advocate for that drug within the company through animal and human trials. The problem was, that by the time that a drug made it to human trials, a "drug champion" might have eight years of his or her career wrapped around this one particular chemical. Not surprisingly, these researchers became eloquent and forceful proponents of "their" compound. So much so that it became difficult to drop drug compounds in early trials. Drugs that never should have made it past animal testing survived until human Phase III clinical trials. By this time, hundreds of millions of dollars of the company's money had been invested, and it took a particularly brave and forceful manager to point out obvious disappointments. A state of corporate cognitive dissonance ensues, group denial. The drug gets pushed until the FDA says no. Most of the oft-quoted $800 million figure for new drugs consists of the amortization of costs for drugs that don't make it through to approval.

Technology can't overcome every obstacle. Even if all the new compounds that chemists could invent were effective drugs, they would still have to be tested on humans – a process that can take at least five years and usually much longer. However much the science might say "this drug will work" based on theoretical grounds and animal testing, the government, rightly, will not approve it until it has been shown to be effective and safe in human trials.

Another great part of the problem with the drug discovery process has been the focus of the major drug companies. Faced with the loss of large moneymakers due to patent expiration, the major drug companies were hungry in the 1990s, and still are today, for large, blockbuster drugs. Drugs that might have helped tens of thousands of people were ignored because of the economic need to find drugs that they could sell to millions of people – usually drugs that were only marginally better than those already sold in well-established markets. When the Cox-2 inhibitor Vioxx was removed from the shelves, many patients simply resorted to ibuprofen, a generic, over-the-counter drug with a good safety profile and a long history.

The approval rate for new drugs by the Food and Drug Administration was stagnant throughout the 1990s and beyond. More depressing still, some of the blockbuster drugs that had been approved, such as Vioxx, had to be pulled from the shelves due to safety concerns.

The personal computer industry, as well, lags dramatically now from its glory years of dramatic growth in the 1980s and 1990s. Every year, it is true, cheaper, more powerful computers become available. But the industry, as a whole, is running up on the hard ground of economics; who actually needs more powerful computers? Aside from a few number-crunchers in the hard sciences and government, the biggest consumers of computer power are video gamers. The Singularity may be stalled if this segment of the population runs out of ready cash.

The Apollo Project is another example of technology with not much follow-through. We reached the moon in the 1960s, but we haven't been back since. Nor has manned flight reached beyond the moon. With all those "Accelerating Returns" you would expect we would be on Mars at least by now.

Although I am a technophile through and through, I must admit to a little skepticism when it comes to the Singularity. For believers in the apotheosis of technology, I would say, beware of the human factor. I don't believe that technology can outrun the ability of people to cope, at least in the global sense (some of us are already overwhelmed). People are ultimately the consumers of technology. Machines, fortunately, don't spend money (yet). And it is money after all, not technology, that makes the world go round.

Regulating Nanotech

Nanotechnology has benefited perversely from some of its critics in the perception that it is a highly advanced, sci-fi sort of technology still way beyond the horizon. This is not necessarily the case. By dollar volume, for instance, one of the biggest use of nanoparticles is in sun cream and cosmetics. The cosmetics maker, L'Oreal, for a while, held the largest number of nanotechnology patents.

However, government regulatory agencies have begun now to focus on nanotechnology. The Better Regulations Taskforce, which advises the government of the United Kingdom has already made some general recommendations. According to the Task Force the Government should:

- enable, through an informed debate, the public to consider the risks for themselves, and help them to make their own decisions by providing suitable information;
- be open about how it makes decisions, and acknowledge where there are uncertainties;

- communicate with, and involve as far as possible, the public in the decision-making process;
- ensure it develops two-way communication channels; and take a strong lead over the handling of any risk issues, particularly information provision and policy implementation.

Glenn Harlan Reynolds, a Tennessee law professor, has written a review for the Pacific Research Institute, called "Forward to the Future: Nanotechnology and Regulatory Policy," perhaps the first serious, non-inflammatory look at the issue in the U.S. [18]. In the review, Reynolds puts forth three potential scenarios for the regulation of nanotechnology in the U.S.:

1. Prohibition.
2. Restriction to the military.
3. Moderate regulation of public use.

Reynolds sees Prohibition as being unworkable, not least because the seeds of the technology are already widely distributed and available. Prohibition would also be wasteful in the benefits to society that would have to be foregone. A military monopoly, Reynolds sees as particularly problematical in that military versions of nanotechnology would likely involve robust weapon systems, and be under the control of Pentagon bureaucrats who are a power unto themselves. He sees as most beneficial a regime of modest regulation emphasizing civilian research and professional responsibility.

"The tools required to develop nanotechnologies are typically small and unobtrusive. The pace of research is accelerating worldwide. Some suggest stopping it, but it is hard to imagine how," says Eric Drexler. "Thus, it seems that this technology, with all its challenges and opportunities, is an unavoidable part of our future." [19]. In this, Drexler seems to be correct.

References

1 Joy, W. Why the Future Doesn't Need Us. *Wired* 8.04, **2003**.
2 The Big Down – Atomtech: Technologies Converging at the Nanoscale, Etc Group, www.etc-group.org/documents/TheBigDown.pdf.
3 Drexler, K.E. From Feynman to Funding. *Bull. Sci. Technol. Soc.* 24: 21–27 (**2004**).
4 Drexler, K.E. *Nanosystems: Molecular Machinery, Manufacturing and Computation.* John Wiley & Sons, **1992**.
5 Arnall, A.H. Future Technologies, Today's Choices, a report for the Greenpeace Environmental Trust, July 2003, www.greenpeace.org.uk.
6 Atwood, M. *Oryx and Crake.* Doubleday, **2003**.
7 Sterling, B. *Holy Fire,* Spectra, **1996**.
8 Malthus, T. *An Essay on the Principle of Population.* Printed for J. Johnson in St. Paul's Churchyard, **1798**.
9 The Royal Society and The Royal Academy of Engineering, *Nanoscience and Nanotechnologies,* **2004**.
10 Connally, P. *Converging Technologies for Improving Human Performance.* Roco, M., Bainbridge, W. (eds.). A report prepared by the National Science Foundation and the Department of Defense.
11 Clynes, M., Kline, N.S. Cyborgs and Space. *Astronautics*: 74–5, September: 26–27 (**1960**).
12 Haraway, D.J. Manifesto for Cyborgs: Science, Technology and Socialist Feminism in the 1980s. *Socialist Review* 65: 108 (**1990**).
13 *Converging Technologies for Improving Human Performance.* Roco, M., Bainbridge, W. (eds.). A report prepared by the National Science Foundation and the Department of Defense.
14 Gibson, W. *Neuromancer.* Ace Books, **1985**.
15 Gsponer, Andre, From the Lab to the Battlefield? Nanotechnology and Fourth-Generation Nuclear Weapons. Disarmament Diplomacy, November, **2002**.
16 Edwards, S.A. *Mind Children,* 21C: Scanning the Future, No. 23.
17 Kurzweil, R. *Law of Accelerating Returns.* www.kurzweilai.net/articles/art0134.html.
18 Reynolds, G.H. Forward to the Future: Nanotechnology and Regulatory Policy. www.pacificresearch.org/puib/sab/techno/forward_to_nanotech.pdf, **2002**.
19 Drexler, K.E. *The Future of Nanotechnology: Molecular Manufacturing.* Essay published on EurekaAlert.org, **2002**.

Chapter 12
Final Thoughts on The Destination

This book has provided a lightning tour of the state of nanotechnology today. We have hit the high points, the big tourist destinations, and visited a few of the interesting out-of-the-way corners. But we can't see it all; indeed, some of it is purposefully hidden from sight. And the field of nanotechnology is growing in all directions at once. If we hoard the few snapshots from our visit and try to revisit the same places a few years hence, the landscape will have become changed beyond recognition. Change, technological and otherwise, is unavoidable.

If – and it's a rather large If – all the development programs mentioned in this book work out as planned, things will be rather different on planet Earth before this century is half over. Our cars will be powered by hydrogen fuel cells; the hydrogen will be generated from water using energy from the sun. Instead of the space shuttle, we will have an elevator into space. Medicine will have been transformed; thousands of diagnostic tests will be run on a few drops of blood in the space of a few minutes. A standard medical test will be the sequencing of your entire DNA. New "smart" cancer drugs will seek out and kill only the cancer cells, without making the treatment seem worse than the disease. Ubiquitous sensors will sample the environment to warn us of dangerous fumes, viruses, or other pathogens. The paralyzed will walk again and the blind will see. Electronics will also be transformed. We will use LEDs instead of incandescent bulbs. Paper thin, wall-sized electronic displays will be everywhere. Electronic components will shrink to nanoscale size and eventually to the size of single molecules. Brain–machine interfaces will allow us to access the contents and abilities of computers with the ease of thought. The personal quantum computer will be thousands of times faster and more powerful than those that clutter our desk-tops today.

I went to college (I hate to admit it) before there were pocket calculators, let alone personal computers. We did our calculations with slide-rules, a primitive instrument that performs multiplication and division by aligning two scales. Operating a slide-rule takes not only a minimum of dexterity but good eyes to see the tiny numbers written on them. During a test in say, physical chemistry, you could hear the "slick, slick" sound of shifting slide rules in the midst of the otherwise dead silence of exam time.

The Nanotech Pioneers. Steven A. Edwards

ISBN: 3-527-31290-0

In March 2005, the MIT Museum accepted a donation of 600 slide-rules from Intellicoat Corp. for their science and technology exhibit. I still own one of these museum pieces.

When I was born in 1951, not only were there no calculators, there were no computers, no Internet, no video games, and no cell phones. Television had been invented, but hardly anyone owned one yet. Space had not yet been penetrated; there were no astronauts. There were cars, but no freeways. Distribution systems were primitive; if you lived in a cold, northern clime, you could not get fresh vegetables during the winter. Biology was still at the descriptive level. DNA had been shown to be the genetic material, but most biochemists did not believe it yet. The structure of the molecule wasn't known. No one worried about whether their food was genetically modified.

When I was three years old I had a brief bout with polio, a year before the Salk vaccine was widely available. I recovered without permanent damage, but others were not so lucky; the girl across the street was permanently paralyzed in one leg. Other childhood diseases that I suffered from included measles, rubella, mumps, and chicken pox. All of these diseases can occasionally be serious and all are just about gone from the American landscape because of childhood vaccination programs.

Some things haven't changed. Just about every Sunday, my wife and I go to a small white church in rural Tennessee that's been standing for 100 years. I lead the singing for Sunday school and take requests from those assembled. We sing great old hymns: *Amazing Grace, The Old Rugged Cross, What a Friend We Have in Jesus*, or *How Great Thou Art*. Some of the lyrics in our songbook were written by Martin Luther or by the old Puritan poet William Cowper. We take communion in a ritual that was started at the Last Supper over 2000 years ago.

A short drive away from us, Rhea County in Tennessee encompasses part of Walden's Ridge, some of the steepest terrain in the U.S. east of the Rockies. Numerous beautiful creeks run off the Ridge toward the Tennessee River through narrow, precipitous gorges that are immune to development. Here, if you are an adept whitewater paddler, you can see a part of Tennessee as it always has been. It was in the Rhea County courthouse that Clarence Darrow memorably fought Williams Jennings Bryant during the Scopes monkey trial over the teaching of Charles Darwin's "Theory of Evolution". The courthouse still stands, and hardly anybody in Rhea County believes in evolution still. Human society is both resistant and accepting of change. Most people exhibit this ambivalence in their own lives.

Most people don't think much about science. But they are happy to use the cell phones, play the video games, send the e-mail, jump on the jet plane and take the antibiotics to get well; they are satisfied with all the fruits of technology that science engenders. They want the light to come on when they turn the switch on a dark night, and become very annoyed if it doesn't.

But change is coming – nanotechnology is both a harbinger and an enabler of overwhelming changes that will follow. When we have our hydrogen cars and our space elevator and our quantum computers, will we have become different people? Will the State and the capitalist system wither away? Will crime and poverty disappear? Will we love our neighbors as ourselves? To all these questions, I reply: It is doubtful. Human nature does not change quickly, if at all.

There are those who seem to expect a cosmic revelation to attend the nanotech revolution. My advice is to forget it. The great Chinese Zen master Joshu was asked, in all seriousness, whether a dog has Buddha-nature or not. "Mu! (nothing)," replied Joshu, relates a famous Zen koan. All sentient beings, including dogs do, in fact, have a Buddha Nature according to Buddhist doctrine. Although the ancient Zen masters did not address this, one assumes that conscious nanotech robots with quantum computer brains would also have a Buddha-nature. But the dog and the sentient robot, like you and I, would still be trapped in the web of karma. That the dog or the robot has Buddha-nature is beside the point. Mu!

Transmission electron micrographs taken of the great hammered blades of old Viking swords reveal nano-structured carbon in the edge of the blades, a case of unconscious nanotechnology employed by ancient artisans for the purpose of making blades sharp. A true nanotech sword would be an even more awesome thing – light as a feather, perhaps built entirely of a single molecule of carbon, with its blade beveled at the edge to the thickness of a single atom. But my guess is that this great nanotech blade would still not cut through the web of karma.

We scientists like to believe that science is value-free – that any moral or ethical problem that crops up because of advancing knowledge is a social construct. Knowledge, by itself, we like to think, is an unqualified good. The old Vikings thought that carbon steel blades were a great invention; the English, however, weren't so sure.

Society, for its part, has always been doubtful about value-free knowledge. For did not God very plainly say "of the tree of the knowledge of good and evil, thou shalt not eat of it: for in the day that thou eatest thereof thou shalt surely die (Genesis 2:17)?" The serpent, subtle creature that he was, stressed the goodies that can come with knowledge: "in the day that ye eat thereof, then your eyes will be opened, and ye shall be as gods ... (Genesis 3:5)." Personally, I have always thought that this was a put up deal. God makes a curious, hairless monkey and then says to him, "Hey, you can have all you want from all these other trees, but don't touch this one tree here." How better to make sure that the monkey did? God may be the original reverse psychologist.

My heretical opinions aside, there is certainly something a little god-like about manipulating matter at its most basic levels. In making the lame to walk and the blind to see. A certain amount of heady grandiosity, much of it warranted, can be perceived in some of the statements of the Nanotech Pioneers.

There is no doubt at all that nanotechnology is very powerful stuff. But remember that it is only the latest chapter in the series of events that started with that first bite of the apple. The history of mankind is also the history of the advancement of knowledge – our cultural evolution. Sometimes knowledge has been used wisely, and sometimes not. But through it all, the good has somehow outweighed the bad and we need to have faith, if not in God, then at least in ourselves, that this will continue. Here we stand, more than six billion of us, 200 000 years from our origin on this Earth, a naked monkey no longer.

Index

The Nanotech Pioneers. Steven A. Edwards

ISBN: 3-527-31290-0

h

i

j

k

l

o